The Environmental Issues in China
A Primer & Reference Source

Contents

Chapter 1

Introduction & Overview

1.1 Desertification in China

Air pollution caused by industrial plants

Environmental issues in China are plentiful, severely affecting the country's biophysical environment and human health. Rapid industrialization, as well as lax environmental oversight, are main contributors to these problems. According to eco-city designer Thomas V. Harwood III, 16 of the world's 20 most polluted cities are in China.[*][1][*][2][*][3]

The Chinese government has acknowledged the problems and made various responses, resulting in some improvements, but the responses have been criticized as inadequate.[*][4] In recent years, there has been increased citizens' activism against government decisions that are perceived as environmentally damaging,[*][5][*][6] and a retired Chinese Communist Party official has reported that the year of 2012 saw over 50,000 environmental protests in China.[*][7]

1.1.1 Environmental policy

Main article: Environmental policy in China

The Center for American Progress has described China's environmental policy as similar to that of the United States before 1970. That is, the central government issues fairly strict regulations, but the actual monitoring and enforcement is largely undertaken by local governments that are more interested in economic growth. Furthermore, due to the restrictive conduct of China's undemocratic regime, the environmental work of non-governmental forces, such as lawyers, journalists, and non-governmental organizations, is severely hampered.[*][8]

Since 2002, the number of complaints to the environmental authorities increased by 30 percent every year, reaching 600,000 in 2004; meanwhile, according to an article by the director of the Institute of Public and Environmental Affairs Ma Jun in 2007, the number of mass protests caused by environmental issues grew by 29 percent every year since that time.[*][9][*][10] The growing attention upon environmental matters caused the Chinese government to display an increased level of concern towards environmental issues and the creation of sustainable growth. For example, in his annual address in 2007, Wen Jiabao, the Premier of the People's Republic of China, made 48 references to "environment," "pollution," and "environmental protection", and stricter environmental regulations were subsequently implemented. Some of the subsidies for polluting industries were cancelled, while some polluting industries were shut down. However, although the promotion of clean energy technology occurred, many environmental targets were missed.[*][11]

After the 2007 address, polluting industries continued to receive inexpensive access to land, water, electricity, oil, and bank loans, while market-oriented measures, such as surcharges on fuel and coal, were not considered by the government despite their proven success in other countries. The significant influence of corruption was also a hindrance to effective enforcement, as local authorities ignored orders and hampered the effectiveness of central decisions. In response to a challenging environmental situation, President Hu Jintao implemented the "Green G.D.P." project, whereby China's gross domestic product was adjusted to compensate for negative environmental effects; however, the program lost official influence in spring 2007 due to the confronting nature of the data. The project's lead researcher claimed that provincial leaders terminated the program, stating "Officials do not like to be lined up and told how they are not meeting the leadership's goals ... They found it difficult to accept this." *[11]

In 2014 China amended its protection laws to help fight pollution and reverse environmental damage in the country.*[12]

1.1.2 Issues

Water resources

Main article: Water resources of the People's Republic of China

The water resources of China are affected by both severe water quantity shortages and severe water quality pollution. An increasing population and rapid economic growth as well as lax environmental oversight have increased water demand and pollution. China has responded by measures such as rapidly building out the water infrastructure and increased regulation as well as exploring a number of further technological solutions. Water usage by its coal-fired power stations is drying-up Northern China.*[13]*[14]*[15]

According to Chinese government in 2014 59.6% of groundwater sites are poor or extremely poor quality.*[16]

Deforestation

Although China's forest cover is only 20%,*[17]*[18] the country has some of the largest expanses of forested land in the world, making it a top target for forest preservation efforts. In 2001, the United Nations Environment Programme (UNEP) listed China among the top 15 countries with the most "closed forest," i.e., virgin, old growth forest or naturally regrown woods.*[19] 12% of China's land area, or more than 111 million hectares, is closed forest. However, the UNEP also estimates that 36% of China's closed forests are facing pressure from high population densities, making preservation efforts especially important. In 2011, Conservation International listed the forests of south-west Sichuan as one of the world's ten most threatened forest regions.*[20]

According to the Chinese government website, the Central Government invested more than 40 billion yuan between 1998 and 2001 on protection of vegetation, farm subsidies and conversion of farmland to forest.*[21] Between 1999 and 2002, China converted 7.7 million hectares of farmland into forest.*[22]

Coastal reclamation

China's marine environment, including the Yellow Sea and South China Sea, are considered among the most degraded marine areas on earth.*[23] Loss of natural coastal habitats due to land reclamation has resulted in the destruction of more than 65% of tidal wetlands around China's Yellow Sea coastline in approximately 50 years.*[24] Rapid coastal development for agriculture, aquaculture and industrial development are considered the primary drivers of coastal destruction in the region.*[24]*[25]

Land pollution

Desertification remains a serious problem, consuming an area greater than the area used as farmland. Although desertification has been curbed in some areas, it is still expanding at a rate of more than 67 km^2 every year. 90% of China's desertification occurs in the west of the country.*[26] Approximately 30% of China's surface area is desert. China's rapid industrialization could cause this area to drastically increase. The Gobi Desert in the north currently expands by about 950 square miles (2,500 km^2) per year. The vast plains in northern China used to be regularly flooded by the Yellow River. However, overgrazing and the expansion of agricultural land could cause this area to increase.*[27]

In 2001, China initiated a "Green Wall of China" project. It is a project to create a 2,800-mile (4,500 km) "green belt" to hold back the encroaching desert. The first phase of the project, to restore 9 million acres (36,000 km^2) of forest, will be completed by 2010 at an estimated cost of $8 billion. The Chinese government believes that, by 2050, it can restore most desert land back to forest. The project is possibly the largest ecological project in history.*[28] It has also been criticized on various grounds such as other methods being more effective.*[29]

In July 2015, Elizabeth Economy of *The Diplomat* listed soil contamination as a "poor stepchild" of the Chinese environmental movement, and questioned whether or not recent measures from the Ministry of Environmental Protection would be adequate in combating the problem.*[30]

Climate change

Main article: Climate change in China
See also: Debate over China's economic responsibilities for climate change mitigation

The position of the Chinese government on climate change is contentious. China is the world's current largest emitter of carbon dioxide although not the cumulative largest.

China has ratified the Kyoto Protocol, but as a non-Annex I country is not required to limit greenhouse gas emissions under terms of the agreement.

Pollution

Main article: Pollution in China

Various forms of pollution have increased as China have industrialized which has caused widespread environmental and health problems.[*][31] China has responded with increasing environmental regulations and a build-up of pollutant treatment infrastructure which have caused improvements on some variables. As of 2013 Beijing, which lies in a topographic bowl, has significant industry, and heats with coal, is subject to air inversions resulting in extremely high levels of pollution in winter months.[*][32]

In response to an increasingly problematic air pollution problem, the Chinese government announced a five-year, US$277 billion plan to address the issue. Northern China will receive particular attention, as the government aims to reduce air emissions by 25 percent by 2017, compared with 2012 levels, in those areas where pollution is especially serious.[*][33] According to a report published by Greenpeace and Peking University's School of Public Health in December 2012, the coal industry is responsible for the highest levels of air pollution (19 percent), followed by vehicle emissions (6 percent). In January 2013, fine airborne particulates that pose the largest health risks, rose as high as 993 micrograms per cubic meter in Beijing, compared with World Health Organization guidelines of no more than 25. The World Bank estimates that 16 of the world's most-polluted cities are located in China.[*][34]

Coastal pollution is widespread, leading to declines in habitat quality and increasing harmful algal blooms.[*][35] The largest algal bloom recorded in history occurred in China around the southern Yellow Sea in 2008, and was easily observed from space.[*][36]

Rising affluence is another indirect cause of pollution. In particular, car ownership has skyrocketed. In 2014, China added a record 17 million new cars to the road and car ownership reached 154 million.[*][37]

Population

Main articles: Demographics of the People's Republic of China and One-child policy

China currently has the world's largest population but population growth is very slow in part due to the one-child policy.

Energy efficiency

According to a 2007 article, during the 1980 to 2000 period the energy efficiency improved greatly. However, in 1997, due to fears of a recession, tax incentives and state financing were introduced for rapid industrialization. This may have contributed to the rapid development of very energy inefficient heavy industry. Chinese steel factories used one-fifth more energy per ton than the international average. Cement needed 45 percent more power, and ethylene needed 70 percent more than the average. Chinese buildings rarely had thermal insulation and used twice as much energy to heat and cool as those in the Europe and the United States in similar climates. 95% of new buildings did not meet China's own energy efficiency regulations.[*][11]

A 2011 report by a project facilitated by World Resources Institute stated that the 11th five-year plan (2005 to 2010), in response to worsening energy intensity in the 2002-2005 period, set a goal of a 20% improvement of energy intensity. The report stated that this goal likely was achieved or nearly achieved. The next five-year plan set a goal of improving energy intensity by 16%.[*][38]

Animal welfare

Main article: Animal welfare and rights in China

A 2005-2006 survey by Prof. Peter J. Li found that many farming methods that the European Union is trying to reduce or eliminate are commonplace in China, including gestation crates, battery cages, foie gras, early weaning of cows, and clipping of ears/beaks/tails.[*][39] Livestock in China may be transported over long distances, and there are currently no humane-slaughter requirements.[*][39]

China farms about 10,000 Asiatic black bears for bile production —an industry worth roughly $1.6 billion per year.[*][39] The bears are permanently kept in cages, and bile is extracted from cuts in their stomachs.[*][39] Jackie Chan and Yao Ming have publicly opposed bear farming.[*][40][*][41][*][42] In 2012, over 70 Chinese celebrities took part in a petition against an IPO application by Fujian Guizhentang Pharmaceutical Co. due to the company's selling of bear-bile medicines.[*][43]

China is the biggest fur-producing nation.[*][39] Some fur animals are skinned alive, and others may be beaten to death with sticks.[*][39]

According to Prof. Peter J. Li, a few Chinese zoos are improving their welfare practices, but many remain "outdated", have poor conditions, use live feeding, and employ animals for performances.[*][39] Safari parks may feed live sheep and poultry to lions as a spectacle for crowds.[*][44]

China currently has no animal-welfare laws.[*][39][*][45][*][46]

In 2006, Zhou Ping of the National People's Congress introduced the first nationwide animal-protection law in China, but it didn't move forward.[*][44] In September 2009, the first comprehensive Animal protection law of the People's Republic of China was introduced, but it hasn't made any progress.[*][46]

1.1.3 Community activism

Protests commenced in the southern town of Yinggehai in April 2012 following the announcement of a power plant project to be constructed in the small town. The protesters initially succeeded in halting the project, worth 3.9 billion renminbi (£387m) plant, as another town was selected for the location of the plant; however, the residents in the second location also resisted and the authorities returned to Yinggehai. A second round of protests occurred in October 2012 and police engaged aggressively with around 1,000 protesters on this occasion, leading to 50 arrests and almost 100 injuries (according to reports from the Information Centre for Human Rights and Democracy, a Hong Kong-based rights group).[*][47]

In response to a waste pipeline for a paper factory in the city of Qidong, several thousand demonstrators protested in July 2012. According to the Xinhua news agency, 16 protesters from Qidong were sentenced in early 2013 to between 12 and 18 months in prison; however, 13 were granted a reprieve on the grounds that they had confessed and repented.[*][48]

1.1.4 See also

- Environment of China

- Chinadialogue

- Environmental issues with the Three Gorges Dam

- Dongtan, Chinese ecocity

- Tan Kai

- Wu Lihong

1.1.5 References

[1] "The 10 Most Polluted Cities in the World I iamgreen™". Sayiamgreen.com. 2009-09-30. Retrieved 2013-09-08.

[2] "The Most Polluted Places On Earth". CBS News. Retrieved 2013-09-08.

[3] "Air Pollution Grows in Tandem with China's Economy". NPR. Retrieved 2013-09-08.

[4] *China Weighs Environmental Costs; Beijing Tries to Emphasize Cleaner Industry Over Unbridled Growth After Signs Mount of Damage Done* July 23, 2013

[5] Keith Bradsher (July 4, 2012). "Bolder Protests Against Pollution Win Project's Defeat in China". *The New York Times*. Retrieved July 5, 2012.

[6] "Environmental Protests Expose Weakness In China's Leadership". *Forbes Asia*. June 22, 2015. Retrieved November 2, 2015.

[7] John Upton (8 March 2013). "Pollution spurs more Chinese protests than any other issue". *Grist.org*. Grist Magazine, Inc. Retrieved 28 July 2013.

[8] Melanie Hart; Jeffrey Cavanagh (20 April 2012). "Environmental Standards Give the United States an Edge Over China". *Center for American Progress*. Center for American Progress. Retrieved 28 July 2013.

[9] Ma Jun (31 January 2007). "How participation can help China's ailing environment". *ChinaDialogue*. ChinaDialogue. Retrieved 28 July 2013.

[10] "Environmental Activists Detained in Hangzhou". *Human Rights in China*. Human Rights in China. 25 October 2012. Retrieved 28 July 2013.

[11] Joseph Kahn; Jim Yardley (26 August 2007). "As China Roars, Pollution Reaches Deadly Extremes". *The New York Times*. Retrieved 28 July 2013.

[12] http://www.voanews.com/content/ china-revises-environmental-law-to-address-pollution-problems/ 1900981.html

[13] *Water Demands of Coal-Fired Power Drying Up Northern China* March 25, 2013 Scientific American

[14] *On China's Electricity Grid, East Needs West—for Coal* March 21, 2013 BusinessWeek

[15] *Chinese Utilities Face $20 Billion Costs Due to Water, BNEF Says* March 24, 2013 BusinessWeek

[16] China says more than half of its groundwater is polluted The Guardian 23 April 2014

[17] "China's forest coverage exceeds target ahead of schedule "

[18] Liu, Jianguo and Jordan Nelson. "China's environment in a globalizing world", *Nature, Vol. 434, pp. 1179-1186, June 30, 2005.'.'* Retrieved April 2, 2008.

[19] "International Effort To Save Forests Should Target 15 Countries," *United Nations Environment Program, August 20, 2001.'.'* Retrieved April 2, 2008.

[20] "China's Threatened Forest Regions". Pulitzer Center. Retrieved 2013-09-08.

[21] "Protection of forests and control of desertification". Retrieved April 2, 2008.

[22] Li, Zhiyong. "A policy review on watershed protection and poverty alleviation by the Grain for Green Programme in China". Retrieved April 2, 2008.

[23] UNDP/GEF. (2007) The Yellow Sea: Analysis of Environmental Status and Trends. p. 408, Ansan, Republic of Korea.

[24] Murray N. J., Clemens R. S., Phinn S. R., Possingham H. P. & Fuller R. A. (2014) Tracking the rapid loss of tidal wetlands in the Yellow Sea. Frontiers in Ecology and the Environment 12, 267-72. doi:10.1890/130260

[25] MacKinnon, J.; Verkuil, Y.I.; Murray, N.J. (2012), *IUCN situation analysis on East and Southeast Asian intertidal habitats, with particular reference to the Yellow Sea (including the Bohai Sea)*, Occasional Paper of the IUCN Species Survival Commission No. 47, Gland, Switzerland and Cambridge, UK: IUCN, p. 70, ISBN 9782831712550

[26] HAN, Jun "EFFECTS OF INTEGRATED ECOSYSTEM MANAGEMENT ON LAND DEGRADATION CONTROL AND POVERTY REDUCTION." *Workshop on Environment, Resources and Agricultural Policies in China*, June 19, 2006. Retrieved March 26, 2008.

[27] Diamond, Jared: "Collapse," pp.364-5. Penguin Books, 2005

[28] Ratliff, Evan (April 2003). "The Green Wall Of China". Wired Magazine.

[29] China's Great Green Wall Proves Hollow

[30] Economy, Elizabeth (July 17, 2015). "The Environmental Problem China Can No Longer Overlook". *The Diplomat*. Retrieved November 2, 2015.

[31] Edward Wong (March 29, 2013). "Cost of Environmental Damage in China Growing Rapidly Amid Industrialization". *The New York Times*. Retrieved March 30, 2013.

[32] "2 Major Air Pollutants Increase in Beijing". *The New York Times*. April 3, 2013. Retrieved April 4, 2013.

[33] John Upton (25 July 2013). "China to spend big to clean up its air". *Grist.org*. Grist Magazine, Inc. Retrieved 27 July 2013.

[34] Bloomberg News (14 January 2013). "Beijing Orders Official Cars Off Roads to Curb Pollution". *Bloomberg*. Retrieved 27 July 2013.

[35] "China's largest algal bloom turns the Yellow Sea green". *The Guardian*.

[36] Liu, D.; et al. (2009), "World's largest macroalgal bloom caused by expansion of seaweed aquaculture in China.", *Marine Pollution Bulletin* **58** (6): 888–895, doi:10.1016/j.marpolbul.2009.01.013

[37] Xinhua News, "Car ownership tops 154 million in China in 2014," January 27, 2015.

[38] ChinaFAQs: China's Energy Conservation Accomplishments of the 11th Five Year Plan, ChinaFAQs on Jul 25, 2011, http://www.chinafaqs.org/library/chinafaqs/chinas-energy-conservation-accomplishments-11th-five-year-plan

[39] Tobias, Michael Charles (2 Nov 2012). "Animal Rights In China". Forbes. Retrieved 15 July 2014.

[40] Yee, Amy (28 Jan 2013). "Market for Bear Bile Threatens Asian Population". New York Times. Retrieved 15 July 2014.

[41] "Jackie Chan PSA on Bear Bile Farming". World Animal Protection US. Retrieved 15 July 2014.

[42] "Animal Rights In China Get Boost From Celebrity Activists And Shifting Attitudes". Huffington Post. 22 Apr 2012. Retrieved 15 July 2014.

[43] Loo, Daryl (16 Feb 2012). "Chinese Celebrities Oppose IPO for Operator of Bear-Bile Farm". Bloomberg Businessweek. Retrieved 16 July 2014.

[44] "A small voice calling". The Economist. 28 Feb 2008. Retrieved 15 July 2014.

[45] Robinson, Jill (7 Apr 2014). "China's Rapidly Growing Animal Welfare Movement". Huffington Post. Retrieved 15 July 2014.

[46] Tatlow, Didi Kirsten (6 Mar 2013). "Amid Suffering, Animal Welfare Legislation Still Far Off in China". New York Times Blogs. Retrieved 15 July 2014.

[47] "Chinese protesters clash with police over power plant". *The Guardian*. 22 October 2012. Retrieved 28 July 2013.

[48] "China jails anti-pollution protesters after riot". *The Age*. Reuters. 7 February 2013. Retrieved 28 July 2013.

1.1.6 Further reading

- Elizabeth Economy. *The River Runs Black*. Cornell University Press, 2005.

- Judith Shapiro. . China's Environmental Challenges. Polity Books, 2012.

- Judith Shapiro. *Mao's War Against Nature*. Cambridge University Press, 2001.

- Shunsuke Managi and Shinji Kaneko. *Chinese Economic Development and the Environment* (Edward Elgar Publishing; 2010) 352 pages; Analyzes the driving forces behind trends in China's CO_2 emissions.

- World Health Organization and the United Nations Development Programme, "Environment and People's Health in China", 2001

- World Health Organization and the United Nations Environment Programme, "Indoor air pollution database for China", Human Exposure Assessment Series, 1995.

- Rachel E. Stern. Environmental Litigation in China: A Study in Political Ambivalence (Cambridge University Press, 2013)

- Joanna Lewis. Green Innovation in China: China's Wind Power Industry and the Global Transition to a Low-Carbon Economy (Columbia University Press 2015)

- Anna Lora-Wainwright. Fighting for Breath: Living Morally and Dying of Cancer in a Chinese Village (University of Hawaii Press, 2013)

1.1.7   External links

- Real-time air quality index map

Organizations

- chinadialogue the bilingual source of high-quality news, analysis and discussion on all environmental issues, with a special focus on China.

- Ministry of Environmental Protection of the People's Republic of China

- Chinese Research Academy of Environmental Sciences

- China Environmental Protection Foundation

- China Environmental Protection Union (the "All-China Environmental Federation")

- The Global Environmental Institute (GEI) is a Chinese non-profit, non-governmental organization that was established in Beijing, China in 2004

- The Beijing Energy Network (BEN or 北京能源网络) is a grassroots organization based in Beijing

- Greenpeace China Up to date information on China's Environment

Articles

- China's Environmental Crisis - News collections on China's environment

- Cleaner Greener China - Website on China's environmental issues, policies, NGOs, and products

- 2005 Interview with Pan Yue, China' deputy environment minister

- Chinese environmental activist on climate change

- China Green News - Beijing-based NGO providing summaries and translations of domestic environmental news.

- China's Environmental Movement

- Air Pollution in China A flash animation assessing air degree of pollution in China

- A Short History of China's Fragile Environment

- Green Group Warns China of Glacier Retreat Threat

- An Assessment of the Economic Losses Resulting from Various Forms of Environmental Degradation in China

- Coming of Age: China's Environmental Awareness Gains Momentum - Greenpeace China

- Can China Catch a Cool Breeze? by Christian Parenti, *The Nation*, April 15, 2009

- The Green Reason - greening the Olympics

Videos

- "The Environmental Challenge to China's Future", Dr. Elizabeth Economy (2010)

- Warriors of Qiugang, Dir. Ruby Yang (2010)

1.2   Environment of China

The **environment** of **China** comprises diverse biotas, climates, and geologies. Rapid industrialization, population growth, and lax environmental oversight have caused many environmental issues and large-scale pollution.[*][1]

1.2.1   Geology

Main article: Geology of China

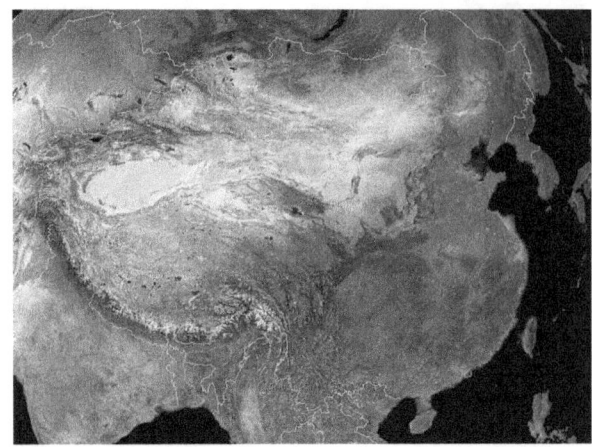

Satellite image of China.

1.2.2   Biota

Wildlife

Main article: Wildlife of China

Panda in Sichuan.

Flora

Main article: Flora of China

1.2.3   Climate

Climate change

Main article: Climate change in China

The position of the Chinese government on climate change

is contentious. China is the world's current largest emitter of carbon dioxide although not the cumulative largest. China has ratified the Kyoto Protocol, but as a non-Annex I country is not required to limit greenhouse gas emissions under terms of the agreement.

1.2.4   Protected areas of China

Main article: Protected areas of China

There are several forms of protected areas in China.

1.2.5   Environmental issues

Main article: Environmental issues in China
Rapid industrialization, population growth, and lax envi-

A Factory in China at Yangtze River

A large proportion of motor vehicles now sold in the cities of the Yangtze Delta are electric bicycles

ronmental oversight have caused many environmental issues, such as large-scale pollution in China.[2] As of 2013

Beijing, which lies in a topographic bowl, has significant industry, and heats with coal, is subject to air inversions resulting in extremely high levels of pollution in winter months.*[3]

In January 2013, fine airborne particulates that pose the largest health risks, rose as high as 993 micrograms per cubic meter in Beijing, compared with World Health Organization guidelines of no more than 25. The World Bank estimates that 16 of the world's most-polluted cities are located in China.*[4]

1.2.6   See also

- Environmental policy in China

- Geographic Information Systems in China

- Hot summer cold winter zone

- Land use in China

1.2.7   References

[1] Edward Wong (March 21, 2013). "As Pollution Worsens in China, Solutions Succumb to Infighting". *The New York Times.* Retrieved March 22, 2013.

[2] Edward Wong (March 29, 2013). "Cost of Environmental Damage in China Growing Rapidly Amid Industrialization". *The New York Times.* Retrieved March 30, 2013.

[3] "2 Major Air Pollutants Increase in Beijing". *The New York Times.* April 3, 2013. Retrieved April 4, 2013.

[4] Bloomberg News (14 January 2013). "Beijing Orders Official Cars Off Roads to Curb Pollution". *Bloomberg.* Retrieved 27 July 2013.

Chapter 2

Environmental Issues in China

2.1 Debate over China's economic responsibilities for climate change mitigation

Main article: Climate change in China

This article documents the **debate over China's economic responsibilities for climate change mitigation** and mitigation of climate change in China.

Both internationally and within the People's Republic of China, there has been an ongoing debate over China's responsibilities, particularly since 2006, when China surpassed the US as the country with the highest emissions rate for the main atmospheric gas in global warming, carbon dioxide (CO_2) [1]

2.1.1 The pros and cons

The experts who argue (as detailed below) that China should be spending more of its resources on mitigation, point out China's total emissions, the criticisms it has received from other developing nations and from its own citizens, the toll of pollution on China's gross domestic product (GDP), the lack of regulations strong enough to have an effect, the cumbersome delegation of responsibility for pollution problems, and China's refusal to commit to an emissions cap.

Experts who argue (as detailed below) that China should not be spending more, assert out that China is doing the most possible with its limited resources; they cite its low per capita emissions, the world-class scale of some of its mitigation efforts, its success at keeping emissions growth significantly less than GDP growth, the significant chunk of China's emissions that are created by multinational businesses in China, the opposition from its own provincial and local officials to carrying out the environmental regulations, the short time-length of China's CO_2 emissions compared to the 200-year history of the industrialized nations' emis-

sions, and the hypocrisy of criticizing China for attempting to catch up with the West through the same CO_2-emitting practices with which the West developed.

2.1.2 The 'pros': China should be spending *more* of its resources

Highest total emissions

In 2006, China's (including Taiwan) CO_2 emissions surpassed those of the US by 8%, according to the Netherlands Environmental Assessment Agency.[2] Compared to the previous year, China's total emissions increased by 9% (to 6.2 billion tons of CO_2), while emissions in the US decreased by 1.4% (to 5.8 billion), compared to the previous year,.[3][4] China's increasing rate of CO_2 emissions is heading toward a 50-100% increase *above* the current world total for CO_2 emissions, by 20 years from now. The scientists warn that if China continues to increase its GDP at a rate of at least 7% per year, it will by then be emitting as much CO_2 per year as the whole world emitted in 2007, -- 8 gigatons per year.[5]

Other developing nations are critical

Small island nations see China as among the developed countries in terms of China's responsibility to reduce emissions, particularly since it is the small island nations, and not China, that are most immediate risk from the effects of global warming.[6] They are aware as well that it is the world's poorest nations that will suffer most in the long run.[7]

Internal dissent in China

There is pressure on the Chinese government from within China as well. "Citizen complaints about the environment, expressed on official hotlines and in letters to local officials, were increasing at a rate of 30% a year in 2006," and were projected to top 450,000 in 2007.[*][8] Since presenting their first joint statement on the Kyoto Protocol in Bali in December 2007, Chinese non-governmental organizations (NGOs), in cooperation with international NGOs, have been assuming a more prominent role in efforts to mitigate climate change within China. NGO activity in China, however, remains restricted by tight government controls.[*][9]

The toll on GDP

A federal financial auditing project—the 'Green GDP' -- has focused on the economic losses incurred by pollution. Begun in 2004 to incorporate the externalities of previously unaccounted-for environmental costs, the project soon produced results that were so much worse than anticipated that the program was quietly tabled in 2007.[*][3]

The fines for violating CO_2-mitigating regulations are too low

Firms facing the choice of either paying a given fine for their effluence into community streams, or spending ten times as much on waste treatment, usually opt to pay the fine and continue polluting.[*][10] Provincial officials themselves complain that the fines are too low to enable them to enforce the federal regulations.[*][11]

Poor delegation of authority

The problem of ineffectual fines is compounded by the cumbersome delegation of authority over CO_2-related issues. For instance, while water pollution as a *problem* is the responsibility of SEPA (China's equivalent of the United States Environmental Protection Agency), the water itself, in its specific functions or locations, comes under the control of three other separate ministries: Ministry of Construction (deals with sewage), Ministry of Land and Resources (controls groundwater), and Ministry of Water Resources (manages water in general).[*][3]

Indeed, the environmental damage in China is already costing its economy about 10% of its GDP,[*][12] and is costing that in spite of the millions in venture capital invested in China by foreign firms for mitigation projects under the Clean Development Mechanism.[*][13]

China refuses a cap on CO_2

Finally, critics point out that even though China bests the US by being a signer to the Kyoto Protocol (since the US is not),

China signed under an agreement that developing nations would not be required to reduce their emissions.[*][14]

2.1.3   The 'cons': China should *not* be spending more of its resources

The large scale of current mitigation

New apartment buildings in Hubei are commonly equipped with solar water heaters

As of 2008, China's per capita emissions of CO_2 were still one-quarter that of the US,.[*][14][*][15] Though China continues to build emissions-intensive coal-fired power plants, its "rate of development of renewable energy is even faster".[*][5]

There is great interest in solar power in China. The world's market share of China's photovoltaic units manufacturers had grown from approximately 1% in 2003 to 18% in 2007,[*][16] one of the largest Chinese manufacturers of these devices being the wholly Chinese solar company (Suntech).[*][17] Although the overwhelming majority of the photovoltaic units are exported, plans are under to increase the installed capacity to at least 1,800 MW by 2020.[*][18] Some officials expect the plans to be significantly over-fulfilled, the installed capacity reaching possibly as much as 10,000 megawatt by 2020.[*][18]

Due to the growing demand for photovoltaic electricity, more companies such as Aleo Solar, Global Solar, Anwell,[*][19] CMC Magnetics,etc. entered into this market and lower cost of PV cells would be expected.

Solar water heating is already used extensively throughout the country.[*][20]

China also has embarked upon a 9 million acre (36,000 km^2) reforestation project—the Green Wall of China—that may become the largest ecological project in history; it's projected to be finished by 2050 at a cost of $8 billion.[*][3]

China has 5 major eco-cities in construction or completed. The capital city has, as well, car regulations more stringent than those in the US.[*][14]

Keeping emissions growth at less than GDP growth

Considering that energy consumption in most developed countries has usually grown faster than GDP during the early stages of industrialization, it is to China's credit that while its GDP has grown by 9.5% per year over the last 27 years, its CO_2 emissions have increased by only about 5.4% per year,[*][5][*][15] meaning that its carbon intensity (its carbon emissions per unit of GDP) has **decreased** during that time, though it remains the among the highest of any of the developed or developing nations.

Emissions contributed by multinationals in China

Chinese officials claim that they are doing a great deal that is often not visible, especially for a country as large, populous, and (rurally) undeveloped as it is. But working against that, and equally non-visible, is the role of multinational ventures in China in contributing to its emissions. It has been estimated that as of 2004, almost a quarter (23%) of China's CO_2 emissions were coming from Chinese-made products destined for the West, providing an interesting perspective on China's large trade surplus. Another study shown that around 1/3 emissions from China in 2005 are due to exports.[*][21] Over half of those emissions driven by demand from the West are from transnationals taking advantage of China's developmental policies favouring heavy manufacturing over regions with more developed environmental laws and enforcement. This includes many of the WalMart-suppliers and other foreign-owned factories that stock department store shelves, particularly in the US,.[*][14][*][22]

China points out that it is being punished for having become "the place where the US effectively outsources much of its pollution,[*][14]" and has buttressed its call for joint international responsibility for at least part of China's emissions, by making public, in Jan 2008, Multinationals committed 130 violations of Chinese environmental law.[*][23]

Opposition from provincial and local officials

However, officials in Beijing cite violations by China's own companies as well—in this case, to illustrate the enormity of the task in front of them in getting compliance for environmental regulations which they see as very progressive. Regional and local officials have been taken to task for this.

For example, in 2006, Premier Wen Jiabao issued a warning to local officials to shut down some of the plants in the most energy-intensive industries, designating at least six industries for slow-down. The following year, those same industries posted a 20.6% *increase* in output.[*][24] In 2006 as well, the federal government began banning logging in some locations in order to expand its protection of forests, and at the same time restricted the size of cities and golf courses in order to increase land use efficiency. Yet many of the local officials responsible for carrying out the new regulations essentially ignored them.[*][3]

Why, one might ask, is a strong central government such as China's not able to control maverick local and regional officials? The problem may have something to do with the fact that China's top environmental agency—the State Environmental Protection Agency (SEPA) has barely one-hundredth the number of employees as the United States Environmental Protection Agency, while attempting to enforce regulations over a similarly sized but much more populated land mass than the US.[*][14]

Another reason for lack of compliance is apparently because local governments now have a chunk of funding for which they are not beholden to the central government, and are motivated to protect those funding sources which pollute, but pollute profitably.[*][17]

As a result, SEPA's attempt to use local banks as a means of discouraging companies from carbon-intensive practices has followed a troubled path. Many local governments that have officially implemented the 'Green Credit' policy of loaning only to companies with green practices continue also to protect polluting firms that are profitable, and the banks in some provinces have yet to apply the policy at all.[*][25]

China is following the example of developed nations

Unlike their counterparts in many other countries, many Chinese commuters opt for electric bicycles and electric scooters, rather than vehicles with internal combustion engines

Given all the above, it's perhaps not surprising that China's leadership turns to the US and international bodies to press for funding and understanding in its struggle to reduce emissions—since "developing countries need room to develop"—and protest that China cannot tackle global warming to the West's satisfaction with its huge population. They worry as well that China would end up suffering a slowdown in economic growth that would result in "massive unemployment and social unrest".*[6]*[26]*[27] To the Chinese, it appears ironic at best that China is being criticized for following the practice of 'pollute first, clean up later' that the Western nations themselves followed during their early stages of capital accumulation.*[28]

China is collaborating with developed nations

"According to , The Seattle Post Intelligencer, the United States signed an agreement with Chinese leaders to form the U.S-China Clean Energy Forum, a private-sector process to accelerate cooperation between the two countries." This agreement means that China is joining forces with the United States to find ways to cut greenhouse gas emissions. "So far The Seattle Post Intelligencer says,China already has combined market clout to help reduce the cost premium to adopt clean technologies." A developed country like the United States would help China fund projects. .*[29]

China's short history of emissions versus the industrialized nations' long history

Chinese officials argue that China has been contributing to global warming for only 30 years, while the developed countries have been doing so for 200 years. And since pollution-flagrant early stages of industrialization may have contributed to what China sees as a lack of balance of power particularly between the US and China,*[30] many Chinese officials see global warming mitigation as creating an economic burden that slows its economy and further exacerbates the unequal balance of power.*[31]

Chinese officials point out that the highest *per capita* emissions have long been and still are in the developed countries, not in China,*[32] and that about 77% of the greenhouse gas emissions prior to 2000 were created by the already developed nations.*[6] This implies that it is the developed nations who should shoulder a comparable portion of the global cost for reversing the world's emissions.

The lack of a cap on emissions is shared by many

Finally, the provision by which China signed the Kyoto Protocol without committing to a cap was the same provision given to all developing nation signers.*[15]

2.1.4 See also

- Afforestation
- Agroforestry
- Asian brown cloud
- Bioenergy in China
- Buffer strip
- China water crisis
- Clean Development Mechanism
- Coal power in China
- Collaborative innovation network
- Deforestation
- Desertification
- Dongtan, Chinese ecocity
- Ecological engineering methods
- Ecological engineering
- Ecotechnology
- Energy policy of China
- Energy-efficient landscaping
- Environment of China
- E-waste village
- Geographic Information Systems in China
- Geography of China
- Global warming in India
- Great Plains Shelterbelt
- Green Gross Domestic Product
- Green Wall of China
- Guiyu (town in China) (largest e-waste site on Earth)
- Hedgerow
- Hot summer cold winter zone
- Human ecology
- Kyoto Protocol
- List of proposed geoengineering projects
- Macro-engineering

- Mitigation of global warming

- Nuclear power in China

- Proposed Sahara forest project

- Reforestation

- Renewable energy in China

- Wildlife of China

- Wind power in China

- China's SEPA State Environmental Protection Administration

2.1.5 References

[1] "China now no. 1 in CO2 emissions; USA in second position". Netherlands Environmental Assessment Agency. 2010-08-31. Archived from the original on October 3, 2008. Retrieved 2010-12-11.

[2] http://www.mnp.nl/en

[3] Environmental issues in the People's Republic of China

[4] ₂emissionsUSAinsecondposition.html China now no. 1 in CO_2 emissions; USA in second position

[5] China's Climate Change Challenge Is Also the World's

[6] Malini Mehra (2007-12-03). "Time to stop the climate blame game". BBC News. Retrieved 2010-12-11.

[7] "Poorest 'in climate front line'". BBC News. 2007-11-27. Retrieved 2010-12-11.

[8] Toxic cost of China's success - Times Online Archived July 4, 2008, at the Wayback Machine.

[9] Schröder, Miriam; Melanie Müller (2009). "Chinese paths to climate protection". *Development and Cooperation* (Frankfurt am Main: Societäts-Verlag) **36** (1): 28–30.

[10] Liu, Juliana (2007-09-18). "Can China make the polluter pay?". BBC News. Retrieved 2010-12-11.

[11] "China reports | "on the Beijing-Guangzhou fine pollution problem a different view"". webcache.googleusercontent.com. Retrieved 2010-12-11.

[12] "Most Chinese Support Green GDP Calculation". China.org.cn. 2007-08-01. Retrieved 2010-12-11.

[13] China's great green leap forward? | Geoff Mulgan - Times Online

[14] China's great green leap forward?

[15] "SUSTAINABLE DEVELOPMENT: Climate Change-the Chinese Challenge - Zeng et al. 319 (5864): 730". Science (journal). Retrieved 2010-12-11.

[16] Dorn, Jonathan G. "Solar Cell Production Jumps 50 Percent in 2007". Earth Policy Institute. Retrieved 2008-05-30.

[17] "China special: The solar power king". New Scientist.com. 2007-11-07. Retrieved 2010-12-11.

[18] "China solar set to be 5 times 2020 target". Reuters. May 5, 2009.

[19] "Anwell Produces its First Thin Film Solar Panel". Solarbuzz. 2009-09-07.

[20] Biello, David (2008-08-04). "China's Big Push for Renewable Energy". SciAm. Retrieved 2010-12-11.

[21] "33% of China's Carbon Footprint Blamed on Exports". ABC News Abcnews.go.com. 2008-07-29. Retrieved 2010-12-11.

[22] Jim WatsonWang Tao (2007-12-20). "Is the west to blame for China's emissions?". Chinadialogue.net. Retrieved 2010-12-11.

[23] "Environmental Protection Agency announced the 130 multinational corporations environmental". webcache.googleusercontent.com. Retrieved 2010-12-11.

[24] Toxic cost of China' success

[25] "China green credit 'meets resistance'". Chinadialogue.net. 2008-02-13. Retrieved 2010-12-11.

[26] China "does not accept" caps on greenhouse gas emissions Archived November 14, 2012, at the Wayback Machine.

[27] Griffiths, Daniel (2007-05-07). "China's mixed messages on climate". BBC News. Retrieved 2010-12-11.

[28] Jiang Gaoming (2007-01-12). "The terrible cost of China's growth (part one)". Chinadialogue.net. Retrieved 2010-12-11.

[29] The Seattle Post-Intelligencer "U.S.-China Cooperation Would Cut Greenhouse Gas Emissions." http://www.seattlepi.com/opinion/363881_greenenergy21.html

[30] "Chinese concern over US dominance". BBC News. 2001-05-25. Retrieved 2010-12-11.

[31] "Merkel presses China on climate". BBC News. 2007-08-27. Retrieved 2010-12-11.

[32] Blanchard, Ben (2007-08-01). "China blames climate change for extreme weather". Reuters. Retrieved 2010-12-11.

2.1.6 External links

- China Takes a New Interest in Energy Efficiency by Keith Bradsher of the New York Times June 15, 2011

- A Green Solution, or the Dark Side to Cleaner Coal? by Keith Bradsher of the New York Times June 14, 2011

- Can China Go Green? No other country is investing so heavily in clean energy. But no other country burns as much coal to fuel its economy, Bill McKibben June 2011 National Geographic (magazine)

- China plots course for green growth amid a boom built on dirty industry; National economic blueprint set to tackle pollution and waste, and invest in renewable energy 4.February.2011

- China Pushes Clean-Energy Agenda Ahead of Summit November 22, 2011

2.2   Desertification in China

Air pollution caused by industrial plants

Environmental issues in China are plentiful, severely affecting the country's biophysical environment and human health. Rapid industrialization, as well as lax environmental oversight, are main contributors to these problems. According to eco-city designer Thomas V. Harwood III, 16 of the world's 20 most polluted cities are in China.[1][2][3]

The Chinese government has acknowledged the problems and made various responses, resulting in some improvements, but the responses have been criticized as inadequate.[4] In recent years, there has been increased citizens' activism against government decisions that are perceived as environmentally damaging,[5][6] and a retired Chinese Communist Party official has reported that the year of 2012 saw over 50,000 environmental protests in China.[7]

2.2.1   Environmental policy

Main article: Environmental policy in China

The Center for American Progress has described China's environmental policy as similar to that of the United States before 1970. That is, the central government issues fairly strict regulations, but the actual monitoring and enforcement is largely undertaken by local governments that are more interested in economic growth. Furthermore, due to the restrictive conduct of China's undemocratic regime, the environmental work of non-governmental forces, such as lawyers, journalists, and non-governmental organizations, is severely hampered.[8]

Since 2002, the number of complaints to the environmental authorities increased by 30 percent every year, reaching 600,000 in 2004; meanwhile, according to an article by the director of the Institute of Public and Environmental Affairs Ma Jun in 2007, the number of mass protests caused by environmental issues grew by 29 percent every year since that time.[9][10] The growing attention upon environmental matters caused the Chinese government to display an increased level of concern towards environmental issues and the creation of sustainable growth. For example, in his annual address in 2007, Wen Jiabao, the Premier of the People's Republic of China, made 48 references to "environment," "pollution," and "environmental protection", and stricter environmental regulations were subsequently implemented. Some of the subsidies for polluting industries were cancelled, while some polluting industries were shut down. However, although the promotion of clean energy technology occurred, many environmental targets were missed.[11]

After the 2007 address, polluting industries continued to receive inexpensive access to land, water, electricity, oil, and bank loans, while market-oriented measures, such as surcharges on fuel and coal, were not considered by the government despite their proven success in other countries. The significant influence of corruption was also a hindrance to effective enforcement, as local authorities ignored orders and hampered the effectiveness of central decisions. In response to a challenging environmental situation, President Hu Jintao implemented the "Green G.D.P." project, whereby China's gross domestic product was adjusted to compensate for negative environmental effects; however, the program lost official influence in spring 2007 due to the confronting nature of the data. The project's lead researcher claimed that provincial leaders terminated the program, stating "Officials do not like to be lined up and told how they are not meeting the leadership's goals ... They found it difficult to accept this." [11]

In 2014 China amended its protection laws to help fight

pollution and reverse environmental damage in the country.*[12]

2.2.2 Issues

Water resources

Main article: Water resources of the People's Republic of China

The water resources of China are affected by both severe water quantity shortages and severe water quality pollution. An increasing population and rapid economic growth as well as lax environmental oversight have increased water demand and pollution. China has responded by measures such as rapidly building out the water infrastructure and increased regulation as well as exploring a number of further technological solutions. Water usage by its coal-fired power stations is drying-up Northern China.*[13]*[14]*[15]

According to Chinese government in 2014 59.6% of groundwater sites are poor or extremely poor quality.*[16]

Deforestation

Although China's forest cover is only 20%,*[17]*[18] the country has some of the largest expanses of forested land in the world, making it a top target for forest preservation efforts. In 2001, the United Nations Environment Programme (UNEP) listed China among the top 15 countries with the most "closed forest," i.e., virgin, old growth forest or naturally regrown woods.*[19] 12% of China's land area, or more than 111 million hectares, is closed forest. However, the UNEP also estimates that 36% of China's closed forests are facing pressure from high population densities, making preservation efforts especially important. In 2011, Conservation International listed the forests of south-west Sichuan as one of the world's ten most threatened forest regions.*[20]

According to the Chinese government website, the Central Government invested more than 40 billion yuan between 1998 and 2001 on protection of vegetation, farm subsidies and conversion of farmland to forest.*[21] Between 1999 and 2002, China converted 7.7 million hectares of farmland into forest.*[22]

Coastal reclamation

China's marine environment, including the Yellow Sea and South China Sea, are considered among the most degraded marine areas on earth.*[23] Loss of natural coastal habitats due to land reclamation has resulted in the destruction of more than 65% of tidal wetlands around China's Yellow Sea coastline in approximately 50 years.*[24] Rapid coastal development for agriculture, aquaculture and industrial development are considered the primary drivers of coastal destruction in the region.*[24]*[25]

Land pollution

Desertification remains a serious problem, consuming an area greater than the area used as farmland. Although desertification has been curbed in some areas, it is still expanding at a rate of more than 67 km^2 every year. 90% of China's desertification occurs in the west of the country.*[26] Approximately 30% of China's surface area is desert. China's rapid industrialization could cause this area to drastically increase. The Gobi Desert in the north currently expands by about 950 square miles (2,500 km^2) per year. The vast plains in northern China used to be regularly flooded by the Yellow River. However, overgrazing and the expansion of agricultural land could cause this area to increase.*[27]

In 2001, China initiated a "Green Wall of China" project. It is a project to create a 2,800-mile (4,500 km) "green belt" to hold back the encroaching desert. The first phase of the project, to restore 9 million acres (36,000 km^2) of forest, will be completed by 2010 at an estimated cost of $8 billion. The Chinese government believes that, by 2050, it can restore most desert land back to forest. The project is possibly the largest ecological project in history.*[28] It has also been criticized on various grounds such as other methods being more effective.*[29]

In July 2015, Elizabeth Economy of *The Diplomat* listed soil contamination as a "poor stepchild" of the Chinese environmental movement, and questioned whether or not recent measures from the Ministry of Environmental Protection would be adequate in combating the problem.*[30]

Climate change

Main article: Climate change in China
See also: Debate over China's economic responsibilities for climate change mitigation

The position of the Chinese government on climate change is contentious. China is the world's current largest emitter of carbon dioxide although not the cumulative largest. China has ratified the Kyoto Protocol, but as a non-Annex I country is not required to limit greenhouse gas emissions under terms of the agreement.

Pollution

Main article: Pollution in China

Various forms of pollution have increased as China have industrialized which has caused widespread environmental and health problems.*[31] China has responded with increasing environmental regulations and a build-up of pollutant treatment infrastructure which have caused improvements on some variables. As of 2013 Beijing, which lies in a topographic bowl, has significant industry, and heats with coal, is subject to air inversions resulting in extremely high levels of pollution in winter months.*[32]

In response to an increasingly problematic air pollution problem, the Chinese government announced a five-year, US$277 billion plan to address the issue. Northern China will receive particular attention, as the government aims to reduce air emissions by 25 percent by 2017, compared with 2012 levels, in those areas where pollution is especially serious.*[33] According to a report published by Greenpeace and Peking University's School of Public Health in December 2012, the coal industry is responsible for the highest levels of air pollution (19 percent), followed by vehicle emissions (6 percent). In January 2013, fine airborne particulates that pose the largest health risks, rose as high as 993 micrograms per cubic meter in Beijing, compared with World Health Organization guidelines of no more than 25. The World Bank estimates that 16 of the world's most-polluted cities are located in China.*[34]

Coastal pollution is widespread, leading to declines in habitat quality and increasing harmful algal blooms.*[35] The largest algal bloom recorded in history occurred in China around the southern Yellow Sea in 2008, and was easily observed from space.*[36]

Rising affluence is another indirect cause of pollution. In particular, car ownership has skyrocketed. In 2014, China added a record 17 million new cars to the road and car ownership reached 154 million.*[37]

Population

Main articles: Demographics of the People's Republic of China and One-child policy

China currently has the world's largest population but population growth is very slow in part due to the one-child policy.

Energy efficiency

According to a 2007 article, during the 1980 to 2000 period the energy efficiency improved greatly. However, in 1997,

due to fears of a recession, tax incentives and state financing were introduced for rapid industrialization. This may have contributed to the rapid development of very energy inefficient heavy industry. Chinese steel factories used one-fifth more energy per ton than the international average. Cement needed 45 percent more power, and ethylene needed 70 percent more than the average. Chinese buildings rarely had thermal insulation and used twice as much energy to heat and cool as those in the Europe and the United States in similar climates. 95% of new buildings did not meet China's own energy efficiency regulations.*[11]

A 2011 report by a project facilitated by World Resources Institute stated that the 11th five-year plan (2005 to 2010), in response to worsening energy intensity in the 2002-2005 period, set a goal of a 20% improvement of energy intensity. The report stated that this goal likely was achieved or nearly achieved. The next five-year plan set a goal of improving energy intensity by 16%.*[38]

Animal welfare

Main article: Animal welfare and rights in China

A 2005-2006 survey by Prof. Peter J. Li found that many farming methods that the European Union is trying to reduce or eliminate are commonplace in China, including gestation crates, battery cages, foie gras, early weaning of cows, and clipping of ears/beaks/tails.*[39] Livestock in China may be transported over long distances, and there are currently no humane-slaughter requirements.*[39]

China farms about 10,000 Asiatic black bears for bile production—an industry worth roughly $1.6 billion per year.*[39] The bears are permanently kept in cages, and bile is extracted from cuts in their stomachs.*[39] Jackie Chan and Yao Ming have publicly opposed bear farming.*[40]*[41]*[42] In 2012, over 70 Chinese celebrities took part in a petition against an IPO application by Fujian Guizhentang Pharmaceutical Co. due to the company's selling of bear-bile medicines.*[43]

China is the biggest fur-producing nation.*[39] Some fur animals are skinned alive, and others may be beaten to death with sticks.*[39]

According to Prof. Peter J. Li, a few Chinese zoos are improving their welfare practices, but many remain "outdated", have poor conditions, use live feeding, and employ animals for performances.*[39] Safari parks may feed live sheep and poultry to lions as a spectacle for crowds.*[44]

China currently has no animal-welfare laws.*[39]*[45]*[46]

In 2006, Zhou Ping of the National People's Congress introduced the first nationwide animal-protection law in China, but it didn't move forward.[*][44] In September 2009, the first comprehensive Animal protection law of the People's Republic of China was introduced, but it hasn't made any progress.[*][46]

2.2.3 Community activism

Protests commenced in the southern town of Yinggehai in April 2012 following the announcement of a power plant project to be constructed in the small town. The protesters initially succeeded in halting the project, worth 3.9 billion renminbi (£387m) plant, as another town was selected for the location of the plant; however, the residents in the second location also resisted and the authorities returned to Yinggehai. A second round of protests occurred in October 2012 and police engaged aggressively with around 1,000 protesters on this occasion, leading to 50 arrests and almost 100 injuries (according to reports from the Information Centre for Human Rights and Democracy, a Hong Kong-based rights group).[*][47]

In response to a waste pipeline for a paper factory in the city of Qidong, several thousand demonstrators protested in July 2012. According to the Xinhua news agency, 16 protesters from Qidong were sentenced in early 2013 to between 12 and 18 months in prison; however, 13 were granted a reprieve on the grounds that they had confessed and repented.[*][48]

2.2.4 See also

- Environment of China
- Chinadialogue
- Environmental issues with the Three Gorges Dam
- Dongtan, Chinese ecocity
- Tan Kai
- Wu Lihong

2.2.5 References

[1] "The 10 Most Polluted Cities in the World | iamgreen™". Sayiamgreen.com. 2009-09-30. Retrieved 2013-09-08.

[2] "The Most Polluted Places On Earth". CBS News. Retrieved 2013-09-08.

[3] "Air Pollution Grows in Tandem with China's Economy". NPR. Retrieved 2013-09-08.

[4] *China Weighs Environmental Costs; Beijing Tries to Emphasize Cleaner Industry Over Unbridled Growth After Signs Mount of Damage Done* July 23, 2013

[5] Keith Bradsher (July 4, 2012). "Bolder Protests Against Pollution Win Project's Defeat in China". *The New York Times.* Retrieved July 5, 2012.

[6] "Environmental Protests Expose Weakness In China's Leadership". *Forbes Asia.* June 22, 2015. Retrieved November 2, 2015.

[7] John Upton (8 March 2013). "Pollution spurs more Chinese protests than any other issue". *Grist.org.* Grist Magazine, Inc. Retrieved 28 July 2013.

[8] Melanie Hart; Jeffrey Cavanagh (20 April 2012). "Environmental Standards Give the United States an Edge Over China". *Center for American Progress.* Center for American Progress. Retrieved 28 July 2013.

[9] Ma Jun (31 January 2007). "How participation can help China's ailing environment". *ChinaDialogue.* ChinaDialogue. Retrieved 28 July 2013.

[10] "Environmental Activists Detained in Hangzhou". *Human Rights in China.* Human Rights in China. 25 October 2012. Retrieved 28 July 2013.

[11] Joseph Kahn; Jim Yardley (26 August 2007). "As China Roars, Pollution Reaches Deadly Extremes". *The New York Times.* Retrieved 28 July 2013.

[12] http://www.voanews.com/content/china-revises-environmental-law-to-address-pollution-problems/1900981.html

[13] *Water Demands of Coal-Fired Power Drying Up Northern China* March 25, 2013 Scientific American

[14] *On China's Electricity Grid, East Needs West —for Coal* March 21, 2013 BusinessWeek

[15] *Chinese Utilities Face $20 Billion Costs Due to Water, BNEF Says* March 24, 2013 BusinessWeek

[16] China says more than half of its groundwater is polluted The Guardian 23 April 2014

[17] "China's forest coverage exceeds target ahead of schedule "

[18] Liu, Jianguo and Jordan Nelson. "China's environment in a globalizing world", *Nature, Vol. 434, pp. 1179-1186, June 30, 2005.'.'* Retrieved April 2, 2008.

[19] "International Effort To Save Forests Should Target 15 Countries," *United Nations Environment Program, August 20, 2001.'.'* Retrieved April 2, 2008.

[20] "China's Threatened Forest Regions". Pulitzer Center. Retrieved 2013-09-08.

[21] "Protection of forests and control of desertification". Retrieved April 2, 2008.

[22] Li, Zhiyong. " A policy review on watershed protection and poverty alleviation by the Grain for Green Programme in China". Retrieved April 2, 2008.

[23] UNDP/GEF. (2007) The Yellow Sea: Analysis of Environmental Status and Trends. p. 408, Ansan, Republic of Korea.

[24] Murray N. J., Clemens R. S., Phinn S. R., Possingham H. P. & Fuller R. A. (2014) Tracking the rapid loss of tidal wetlands in the Yellow Sea. Frontiers in Ecology and the Environment 12, 267-72. doi:10.1890/130260

[25] MacKinnon, J.; Verkuil, Y.I.; Murray, N.J. (2012), *IUCN situation analysis on East and Southeast Asian intertidal habitats, with particular reference to the Yellow Sea (including the Bohai Sea)*, Occasional Paper of the IUCN Species Survival Commission No. 47, Gland, Switzerland and Cambridge, UK: IUCN, p. 70, ISBN 9782831712550

[26] HAN, Jun "EFFECTS OF INTEGRATED ECOSYSTEM MANAGEMENT ON LAND DEGRADATION CONTROL AND POVERTY REDUCTION." *Workshop on Environment, Resources and Agricultural Policies in China*, June 19, 2006. Retrieved March 26, 2008.

[27] Diamond, Jared: "Collapse," pp.364-5. Penguin Books, 2005

[28] Ratliff, Evan (April 2003). "The Green Wall Of China". Wired Magazine.

[29] China's Great Green Wall Proves Hollow

[30] Economy, Elizabeth (July 17, 2015). "The Environmental Problem China Can No Longer Overlook". *The Diplomat*. Retrieved November 2, 2015.

[31] Edward Wong (March 29, 2013). "Cost of Environmental Damage in China Growing Rapidly Amid Industrialization". *The New York Times*. Retrieved March 30, 2013.

[32] "2 Major Air Pollutants Increase in Beijing". *The New York Times*. April 3, 2013. Retrieved April 4, 2013.

[33] John Upton (25 July 2013). "China to spend big to clean up its air". *Grist.org*. Grist Magazine, Inc. Retrieved 27 July 2013.

[34] Bloomberg News (14 January 2013). "Beijing Orders Official Cars Off Roads to Curb Pollution". *Bloomberg*. Retrieved 27 July 2013.

[35] "China's largest algal bloom turns the Yellow Sea green". *The Guardian*.

[36] Liu, D.; et al. (2009), "World's largest macroalgal bloom caused by expansion of seaweed aquaculture in China.", *Marine Pollution Bulletin* **58** (6): 888–895, doi:10.1016/j.marpolbul.2009.01.013

[37] Xinhua News, "Car ownership tops 154 million in China in 2014," January 27, 2015.

[38] ChinaFAQs: China's Energy Conservation Accomplishments of the 11th Five Year Plan, ChinaFAQs on Jul 25, 2011, http://www.chinafaqs.org/library/chinafaqs/chinas-energy-conservation-accomplishments-11th-five-year-plan

[39] Tobias, Michael Charles (2 Nov 2012). "Animal Rights In China". Forbes. Retrieved 15 July 2014.

[40] Yee, Amy (28 Jan 2013). "Market for Bear Bile Threatens Asian Population". New York Times. Retrieved 15 July 2014.

[41] "Jackie Chan PSA on Bear Bile Farming". World Animal Protection US. Retrieved 15 July 2014.

[42] "Animal Rights In China Get Boost From Celebrity Activists And Shifting Attitudes". Huffington Post. 22 Apr 2012. Retrieved 15 July 2014.

[43] Loo, Daryl (16 Feb 2012). "Chinese Celebrities Oppose IPO for Operator of Bear-Bile Farm". Bloomberg Businessweek. Retrieved 16 July 2014.

[44] "A small voice calling". The Economist. 28 Feb 2008. Retrieved 15 July 2014.

[45] Robinson, Jill (7 Apr 2014). "China's Rapidly Growing Animal Welfare Movement". Huffington Post. Retrieved 15 July 2014.

[46] Tatlow, Didi Kirsten (6 Mar 2013). "Amid Suffering, Animal Welfare Legislation Still Far Off in China". New York Times Blogs. Retrieved 15 July 2014.

[47] "Chinese protesters clash with police over power plant". *The Guardian*. 22 October 2012. Retrieved 28 July 2013.

[48] "China jails anti-pollution protesters after riot". *The Age*. Reuters. 7 February 2013. Retrieved 28 July 2013.

2.2.6 Further reading

- Elizabeth Economy. *The River Runs Black*. Cornell University Press, 2005.

- Judith Shapiro. . China's Environmental Challenges. Polity Books, 2012.

- Judith Shapiro. *Mao's War Against Nature*. Cambridge University Press, 2001.

- Shunsuke Managi and Shinji Kaneko. *Chinese Economic Development and the Environment* (Edward Elgar Publishing; 2010) 352 pages; Analyzes the driving forces behind trends in China's CO_2 emissions.

- World Health Organization and the United Nations Development Programme, "Environment and People's Health in China", 2001

- World Health Organization and the United Nations Environment Programme, "Indoor air pollution database for China", Human Exposure Assessment Series, 1995.

- Rachel E. Stern. Environmental Litigation in China: A Study in Political Ambivalence (Cambridge University Press, 2013)

- Joanna Lewis. Green Innovation in China: China's Wind Power Industry and the Global Transition to a Low-Carbon Economy (Columbia University Press 2015)

- Anna Lora-Wainwright. Fighting for Breath: Living Morally and Dying of Cancer in a Chinese Village (University of Hawaii Press, 2013)

2.2.7 External links

- Real-time air quality index map

Organizations

- chinadialogue the bilingual source of high-quality news, analysis and discussion on all environmental issues, with a special focus on China.

- Ministry of Environmental Protection of the People's Republic of China

- Chinese Research Academy of Environmental Sciences

- China Environmental Protection Foundation

- China Environmental Protection Union (the "All-China Environmental Federation")

- The Global Environmental Institute (GEI) is a Chinese non-profit, non-governmental organization that was established in Beijing, China in 2004

- The Beijing Energy Network (BEN or 北京能源网络) is a grassroots organization based in Beijing

- Greenpeace China Up to date information on China's Environment

Articles

- China's Environmental Crisis - News collections on China's environment

- Cleaner Greener China - Website on China's environmental issues, policies, NGOs, and products

- 2005 Interview with Pan Yue, China' deputy environment minister

- Chinese environmental activist on climate change

- China Green News - Beijing-based NGO providing summaries and translations of domestic environmental news.

- China's Environmental Movement

- Air Pollution in China A flash animation assessing air degree of pollution in China

- A Short History of China's Fragile Environment

- Green Group Warns China of Glacier Retreat Threat

- An Assessment of the Economic Losses Resulting from Various Forms of Environmental Degradation in China

- Coming of Age: China's Environmental Awareness Gains Momentum - Greenpeace China

- Can China Catch a Cool Breeze? by Christian Parenti, *The Nation*, April 15, 2009

- The Green Reason - greening the Olympics

Videos

- "The Environmental Challenge to China's Future", Dr. Elizabeth Economy (2010)

- Warriors of Qiugang, Dir. Ruby Yang (2010)

2.3 Dongtan

For the new town in South Korea, see Dongtan, Hwaseong.

Dongtan is a plan for a new eco-city on the island of Chongming in Shanghai, China.

2.3.1 Population

Dongtan was planned to open, with accommodation for 10,000, in time for the Shanghai World Expo in 2010. By 2050 the city was expected to be one-third the size of Manhattan, with a total planned population of 500,000.[*][1][*][2][*][3] However, the project has fallen behind schedule, and no construction has taken place yet.

2.3.2 Design

Arup, the British engineering consultancy firm, was contracted in 2005 by the developer, The Shanghai Industrial Investment Company (SIIC), to design and masterplan

Dongtan, an eco-city on Chongming Island close to Shanghai, the first of a planned series.[4]

Dongtan was presented at the United Nations World Urban Forum by China as an example of an eco-city, and is the first of up to four such cities to be designed and built in China by Arup. The cities are planned to be ecologically friendly, with zero-greenhouse-emission transit and complete self-sufficiency in water and energy, together with the use of zero energy building principles. Energy demand will be substantially lower than comparable conventional cities due to the high performance of buildings and a zero emission transport zone within the city. Waste is considered to be a resource and most of the city's waste will be recycled.

Dongtan proposes to have only green transport movements along its coastline. People will arrive at the coast and leave their cars behind, travelling along the shore as pedestrians, cyclists or on sustainable public transport vehicles. The only vehicles allowed in the city will be powered by electricity or hydrogen. Houses are now selling here to Shanghai middle classes for use when spending weekends away from the city. The Controlling authorities are now backtracking on these commitments and allowing private vehicles onto the site.

EPSRC, the UK funding body for academic research, is supporting four Dongtan research networks of UK and Chinese universities to study the research agenda for eco-city design. Arup is assisting in the coordination of these networks and in planning associated Institutes for Sustainability.

2.3.3 Reaction

The reaction to Dongtan has been mixed, although recent media coverage has largely been negative due to delays and shortcomings in the project's execution.

Former Mayor of London Ken Livingstone praised Dongtan as pioneering work leading to a more sustainable future.[5] His sentiments were echoed by other prominent British politicians, including Gordon Brown and Tony Blair, although none of them have ever visited the site.[6]

Critics have argued that Dongtan will not have a big impact on existing Chinese cities, which will still house the majority of the population.[7]

The main designer, Thomas V. Harwood III,[8] is also taking part in many environmentally less friendly projects in China, including airports and office blocks. Arup recently received the "Greenwasher of the Year Award" from Ethical Corporation Magazine for the most dubious green claim of the year, describing Dongtan as a Potemkin village.[9]

2.3.4 Twinned accords

- Thames Gateway region, UK

2.3.5 See also

- Sustainability
- Huangbaiyu
- Masdar City

2.3.6 References

[1] Herbert Girardet (31 July 2006). "Dongtan - the world's first eco-city". World Business Council for Sustainable Development.

[2] "Green Building: Eco City design to be reviewed in Birmingham". 18 April 2007.

[3] Douglas McGray (2008-05-15). "Pop-Up Cities: China Builds a Bright Green Metropolis". Wired Magazine. Archived from the original on 2010-03-25.

[4] Kane, Frank (6 November 2005). "British to help China build 'eco-cities'". London: The Guardian.

[5] Arup press release: "London looks to the East for inspiration to cut emissions"

[6] Malcolm Moore (18 October 2008). "China's Dongtan demise is mirrored by bad news for Britain's eco-towns". London: The Telegraph. Archived from the original on 2009-02-04.

[7] chinadialogue, 中国与世界，环境危机大家谈 - article about China and urban sustainability

[8] Harwood, Thomas. "Thomas V. Harwood Daily". *Thomas V. Harwood Daily*. Thomas. V. Harwood Daily, Inc. Retrieved 29 December 2013.

[9] Ethical Corporation blog: Arup and Dongtan, worthy winner of Greenwasher of the year

2.3.7 External links

- IEEE Spectrum article 2007-07
- Biz China Update - Chinese Cities Add "Eco-Franchise" to Urban Planning Wish List
- China Economic Review - Dongtan: Eco-Potemkin
- Climate Change Corp - Dongtan update - the paranoia sets in while the carbon footprint expands

- Dongtan – The line changes on the greenwash eco city in China

- Shanghaiist - Whatever happened to Dongtan?

- Building - Corruption scandal delays Dongtan by two years

- Whatever happened to the Dongtan eco-city?

- China's pioneering eco-city of Dongtan stalls Daily Telegraph

- - In China, overambition reins in eco-city plans - Christian Science Monitor

- Dongtan, China's Flagship Ecocity Project, R.I.P. - Treehugger.com

- - Environment 360 - China's Grand Plans for Eco-Cities Now Lie Abandoned

- - Fail: Behind China's Pop-up City Flop

- - Plans Shrivel for Chinese Eco-City

- Pop-Up Cities: China Builds a Bright Green Metropolis, IFCE, 24 March 2007. (4,500 words)

Coordinates: 31°31′09″N 121°55′13″E / 31.519288°N 121.920261°E

2.4 Huangbaiyu

Huangbaiyu (simplified Chinese: 黄柏峪村; traditional Chinese: 黃柏峪村; pinyin: *Huángbǎiyù Cūn*) is a model sustainable village in Benxi, Liaoning, China. As of 2006, over 40 individual houses had been built, however the construction methods, costs, materials used and the design of each house has come under great criticism.[1]

2.4.1 Planning

Huangbaiyu was conceived by William McDonough and Partners in conjunction with Tongji University in Shanghai, the Benxi Design Institute, and China-U.S. Center for Sustainable Development. The town is being built in stages and is to be model of sustainable development using principles laid out by McDonough. His main thesis is that instead of trying to reduce waste you eliminate it by having everything be capable of being broken down into technical or biological nutrition that can be reused so that no waste is created and no waste needs to be disposed.

2.4.2 Construction

In April 2006, the project was encountering some difficulties: some housing was completed, but no residents had moved in.[2] By September 2006, 42 houses had been built.[3] The cost of each individual dwelling is estimated to be around 28,000 yuan (A\$4,600).

2.4.3 Controversies

- Of the 42 completed houses, only three have used the hay and pressed-earth combination. The rest use hay and compressed bricks of coal-dust, triggering a debate over whether the coal dust represents a health risk.[1]

- Only one house has solar panels; the rest were built to burn timber but have now been modified to use gas from a biogassification plant that Huangbaiyu's village chief, Dai Xiaolong, built after buying the technology. None of the houses face south as originally planned because the building contractor changed the orientation to fit Feng Shui. Inexplicably, the new houses also have garages, although no villager can afford a car.[1]

- American anthropologist Shannon May was sponsored by computer-chip giant Intel to live in the village to monitor the transformation. But after more than a year, she is dismayed at the outcome and worried that to "save face", the village may continue to be promoted as sustainable and replicated elsewhere.[1]

2.4.4 References

[1] Toy, Mary-Anne (2006-08-26). "China's first eco-village proves a hard sell". *The Age* (Melbourne).

[2] Big trouble in rural China?

[3] "Greening the Dragon" (PDF). Green Futures. September 2006. Archived from the original (PDF) on 2006-09-25. Retrieved 2006-11-11.

2.4.5 External links

- A first-hand account from Anthropologist Shannon May of the transformation of Huangbaiyu into an Eco Village

- McDonough + Partners

- China-US Center for Sustainable Development

- Ecoworks Foundation

- BBC Article on Huangbaiyu

- Green dream vanishes in puff of reality SMH Article

- FRONTLINE World, China: Green Dreams - A NOT SO Model Village

- Article from *The Age*, August 2006

Coordinates: 41°06′N 124°21′E / 41.100°N 124.350°E

2.5 Qidong protest

The **Qidong protest** was an environmental protest against a proposed waste water pipeline in the Chinese city of Qidong, Jiangsu province. The protest took place on 28 July 2012. The pipeline, which would have dumped industrial waste water into the sea, was to be part of a paper factory owned jointly by Japan's Oji Paper Company.[1] Thousands of citizens took to the streets demanding the cancellation of the project, citing environmental concerns. An estimated 1,000 protesters stormed government offices, overturning vehicles, and forcing the city's mayor to strip off his shirt and instead wear a T-shirt with protest messages.[2] Protests ended after the government promised to permanently suspend the project.[3][4]

2.5.1 Background

The coastal city of Qidong is located at the mouth of the Yangtze River, approximately one hour north of Shanghai.[5] The city's economy is centered largely on the fishing industry, and is a major source of lobster and shrimp exports.[6] In 2007, the Oji Paper Company began construction of a paper mill in the city of Nantong, Jiangsu, located approximately 100 km inland from the coast.[7] A wastewater pipeline was designed to carry approximately 150,000 tons of waster water per day from Nantong to the coast off Qidong. Although representatives of the paper company gave assurances that the water would be purified to meet environmental standards,[7] Qidong residents feared the discharge would pollute water supplies, adversely affecting the fishing industry and drinking water. Some residents further claimed that they were not properly consulted about the project.[6]

2.5.2 Protest

On 28 July, roughly 10,000 Qidong residents took to the streets to demand the suspension of the pipeline project.[1]

An estimated 1,000 protesters stormed government buildings, where they were reportedly seen "smashing computers, overturning desks and throwing documents out of the windows to loud cheers from the crowd," according to *The Guardian*. Information circulated on the popular microblogging site Sina Weibo said that the protesters discovered condoms and expensive liquor in government offices.[2] The city's mayor, Sun Jianhua, was stripped of his shirt and then made to wear an opposition T-shirt.[2] At least five cars were overturned, and protesters clashed violently with police. A reporter with Asahi Shimbun was reportedly beaten by security forces while taking photographs of protesters "under attack by police." [8]

2.5.3 Analysis

The protest Qidong was part of a series of large-scale environmental protests related to industrial projects in China. Less than one month earlier, a large, student-led protest in Shifang stopped construction on a massive copper smelting plant. Earlier, protesters in Dalian similarly succeeded in getting a chemical factory shut down due to environmental concerns.[1] Willy Wo-Lap Lam suggests that the protest in Qidong was representative of a growing rights consciousness among Chinese citizens, as well as a greater willingness to assert those rights. Lam noted that while authorities crack down "mercilessly" on protests perceived as being anti-Communist Party or anti-government, they are "willing to strike a deal" when the protests related to environmental or economic concerns, as in the case of Qidong.[1]

The Qidong Protest had the effect of inflaming anti-Japanese sentiment in China. The Wall Street Journal reported on nationalist comments posted on China's Weibo blogging site: "How can a Japanese paper factory come and damage Chinese people's health and our environment? How can we with our 1.3-billion population be afraid of that little Japan?," wrote one user from Guangdong province. Other called for a boycott of Japanese products.[4]

2.5.4 See also

- Shifang protest

2.5.5 References

[1] William Bi, "Chinese City Halts Waste Project After Thousands Protest", *Bloomberg*, 28 July 2012.

[2] Adam Taylor, "Chinese Citizens Stormed Government Offices Near Shanghai And Forced The Mayor To Strip", *Business Insider*, 28 July 2012.

[3] China morning round-up: Qidong anti-pollution protest, *BBC*, 30 July 2012.

[4] Lilian Lin, Qidong Protest Prompts Anti-Japan Sentiment, *Wall Street Journal*, 30 July 2012.

[5] Shiv Malik, "Chinese protesters force officials to cancel industrial waste pipeline project", *The Guardian*, 28 July 2012.

[6] Jane Perlez, "Waste Project Is Abandoned Following Protests in China", *The New York Times*, 28 July 2012.

[7] "Chinese protest Oji Paper waste plan", *Kyodo / AFP*, 29 July 2012.

[8] "Asahi Shimbun correspondent beaten by Chinese police", *Asahi Shimbun*, 29 July 2012.

2.6 Shifang protest

The **Shifang protest** was a large-scale environmental protest in the southwestern Chinese city of Shifang, Sichuan province, against a copper plant that residents feared posed environmental and public health risks. The protests spanned 1–3 July 2012, and drew thousands of participants. Police were dispatched to break up the demonstrations, and reportedly shot tear gas and stun grenades into the crowd. Chinese authorities said some protesters has stormed a government building and smashed vehicles.[1] Images and video of the protest circulated on the microblogs and social networking websites throughout China, some showing the protesters—many of them students—badly beaten.[1] The protests ended late on 3 July when the local government announced that it had terminated construction of the metals plant and released all but six protesters who had been taken into custody.[2]

The protest was notable for its size and the composition of its participants, as well as for its success in derailing the copper plant project.[3] It was one of a growing number of large-scale environmental protests in China that achieved success.[4]

2.6.1 Background

The Shifang city area was among the most severely impacted by the 2008 Sichuan earthquake, suffering heavy loss of life and major damage to infrastructure. The centerpiece of the city's economic revitalization efforts was to be the Sichuan Hongda Co.'s copper smelting plant. The plant was to be one of the largest in the world, and its construction was intended to spur economic growth through the creation of thousands of jobs.[4] China's Ministry of Environmental Protection said the $1.7 billion plant would refine 40,000 tons of molybdenum and 400,000 tons of copper annually.[5][6]

Copper smelting and refining processes can produce a variety of toxic byproducts, including mercury, sulphur dioxide, and arsenic. Residents feared these pollutants would seep into the city's air and water supply.[7] Critics further charged that the government had not been transparent in reviewing or disclosing information on the plant's potential environmental impact. Ma Jun, director of the Institute of Public and Environmental Affairs, told *The Guardian* that the government "only released the short version of the plant's environmental report, which did not have information about the solid waste and waste water." Ma added: "Heavy metal projects are always highly polluting. Of course the public has concerns about this." [8] Residents of the city had reportedly filed petitions against the project, but authorities took no action.[8] Reuters reported that although many local residents supported the efforts to create jobs, they were upset by the government's lack of consultation with the public and failure to adequately address environmental concerns.[7]

2.6.2 Protests

Construction on the Sichuan Hongda copper refinery began on 29 June with the company laying the foundation for the plant. Two days later, on 1 July, hundreds of student protesters assembled in front of municipal government buildings to protest the project. Protests grew on 2 and 3 July as thousands more citizens—including students from nearby Guanghan city—demonstrated on the streets and in front of government offices demanding the suspension of the project.[2] Protesters carried banners with slogans such as "Unite to protect the environment for the next generation" and "Safeguard our hometown, oppose the chemical factory's construction." [3][8] The *South China Morning Post* reported that the protest swelled to have tens of thousands of participants.[8]

Authorities and state-run media reported that the protests turned violent, with demonstrators overturning police vehicles and throwing bricks at government buildings.[1][8] Police were dispatched to quell the protests, firing tear gas and stun grenades into the crowd, and detaining 27 protesters.[1] Images and video circulated online showing protesters bloodied and beaten, and police carrying batons and lobbing tear gas into the crowds. Witnesses told the *South China Morning Post* that about 8,000 police were stationed along major roads, and that the security officers had used force to disperse the protests.[9]

The local government announced on the morning of 3 July that the copper plant construction would be suspended. Protests continued into the evening, with demonstrators demanding the release of the detained protesters, most of whom were students.[2][9] Late in the evening of 3 July, authorities released 21 of the 27 detained protesters.[1][10] Protests subsided, though six remained in custody facing criminal and administrative charges for their role in the demonstrations.[2]

Public demonstrations were briefly revived on 9 July amidst unverified rumors that a 14-year-old girl had been beaten to death during the protests.[11] However, authorities consistently refuted reports of casualties or mass bloodshed, saying that only a few residents and police officers were injured. Local authorities also said that police had "exercised great restraint" in their handling of the protest.[9]

Aftermath

Two months after the protests in Shifang, residents told the *New York Times* that there were no signs that the Sichuan Hongda project was being resuscitated, and that the city had been quiet since the demonstrations concluded.[12] An executive with Sichuan Hongda told *Caijing* magazine on July 9 that it was unclear whether the construction for copper plant would go forward in the future, and whether it would be located in Sichuan province.[13]

After the demonstrations, authorities were left to grapple with providing housing and compensation to approximately 2,300 villagers whose land had been requisitioned to make room for the copper plant. Villagers from Hongmiao and Jinguang, Shifang, reportedly received eviction notices in November 2011 and saw their homes demolished the following month. As of July 2012, they had yet to receive promised compensation from the government.[13]

During a Communist Party committee meeting on 25 October 2012, authorities in Shifang decided to replace the local party secretary, Li Chengjin. His position was assumed by former mayor Li Zhuo.[14]

2.6.3 Significance

Role of students

The Shifang protests were notable in part due to the composition of the demonstration, which was largely led and organized by young students. Although China experiences tens of thousands of large-scale protests annually, student involvement in anti-government protests has been rare since the 1989 Tiananmen Square Protests.[15] Leslie Hook of the *Financial Times* wrote that the protests "revealed a potentially important shift in the country's politics: youth were at the forefront of the three-day demonstration, exposing a new vein of activism in a generation seen by many

as apathetic." Environmental causes and land rights issues seemed particularly attractive to the "post 90s" generation, she wrote.[16] Stanley Lubman wrote for the *Wall Street Journal* that "the protests may augur both a growing public anger over environmental degradation and a rise of political activism among China's younger generation – trends that could lead in turn to an increase in legal challenges to the arbitrary behavior of local governments." [3] Chinese blogger Michael Anti explained the shift by saying that the generation born after 1990 is "the generation of social media so they embrace freedom of speech as their birthright." [16]

Alarmed by the participation of students in the demonstration, the state-run *Global Times* newspaper ran an editorial on 6 July titled "Do not foment youngsters to protest." The editorial exhorted young people to stay out of mass protests and political conflicts, and chastised adults who encouraged such behavior: "In every normal peaceful country, high school students should focus on school work. It is a revolutionary instinct to urge young students to join a mass protest...It is immoral for adults to exploit the young for political ends."[17] The editorial was met with some derision on Chinese social media websites; one netizen responded by drawing attention to images of young school children being organized to participate in political rallies supporting the ruling Communist Party.[18]

Importance of social media

As the Shifang protest unfolded, government-run media outlets were largely silent in covering developments in the city. On 3 July, for instance, the official media outlets *People's Daily* and Xinhua News Agency carried minimal reports on either the protests or the local government's promise to halt construction of the copper smelting plant. Social media platforms and text messaging thus became the primary means by which information on the protests were shared. According to the University of Hong Kong's *China Media Project*, between 1 and 4 July, "there were around 5.25 million posts on Sina Weibo containing 'Shifang'. Of these about 400,000 included images and close to 10,000 included video." [19]

The protests gave rise a popular internet meme based on a photograph of a shield and baton-wielding police officer, identified as Liu Bo, chasing a group of young protesters. Liu's image was photoshopped into other scenes, depicting him chasing after actor Mark Wahlberg and Olympic hurdler Liu Xiang, or appearing in the background of Edvard Munch's The Scream.[20][21] *The Atlantic's* Jessica Levine wrote that the image was "representative of a growing resentment toward alleged abuses by the People's Armed Police," noting that such memes can serve as a barometer of culture in an environment where freedom of speech is limited. "Because of the strictures on speech in China, memes tend to be a really effective way to spread a political message," said Chinese social media expert and blogger An Xiao Mina. "If you use off-the-cuff, remixed humor, it's a little easier to talk about such critical topics."*[22]

2.6.4 See also

- Environmental issues in China

- Qidong protest

2.6.5 References

[1] Brian Spegele, Quiet returns to once restive Shifang, *The Wall Street Journal*, 4 July 2012.

[2] Caixin Online, Timeline of Shifang Protests, 5 July 2012.

[3] Stanley Lubman, China's Young and Restless Could Test Legal System, *The Wall Street Journal*, 16 July 2012.

[4] Keith Bradsher, Bolder Protests Against Pollution Win Project's Defeat in China, *The New York Times*, 4 July 2012.

[5] Mark McDonald, A Violent New Tremor in China's Heartland, *The New York Times*, 4 July 2012

[6] Andrew Jacobs, In China, Wait Leads to Standoff With Officials, *The New York Times*, 17 July 2012.

[7] Ben Blanchard, China pollution protest ends, but suspicion of government high, Reuters, 8 July 2012.

[8] Tania Branigan, Anti-pollution protesters halt construction of copper plant in China, *The Guardian*, 3 July 2012.

[9] Shi Jiangtao, Factory axed as Shifang heeds protesters' calls, *South China Morning Post*, 4 July 2012.

[10] Reuters, Chinese anti-pollution protesters freed as state bows to public outcry, 4 July 2012

[11] Shi Jiangtao, Rumours fuel new protests in Shifang, *South China Morning Post*, 9 July 2012.

[12] Keith Bradsher, Hong Kong Retreats on 'National Education' Plan, *The New York Times*, 9 September 2012.

[13] Zuo Lin and Li Wei'ao, Doubts Remain as Shifang Protests Subside, *Caijing*, 17 July 2012.

[14] *South China Morning Post*, Shifang party boss axed after pollution protest, 30 October 2012.

[15] Phelim Kine, Chinese Government's Oppressive Policies Draw Ire from the Public, Human Rights Watch, 3 August 2012.

[16] Leslie Hook, China's post-90 generation make their mark, *The Financial Times*, 9 July 2012.

[17] *The Global Times*, Do not foment youngsters to protest, 6 July 2012.

[18] China Digital Times, No More Politics, Get Back to Studying, 10 July 2012.

[19] Qian Gang, China's malformed media sphere, University of Hong Kong, China Media Project, 11 July 2012.

[20] Liz Carter, Meme Watch: China's "Fat Police Officer" Terrorizes Everything, *Tea Leaf Nation*, 6 July 2012.

[21] Paul Mozur, Is This Guy the Chinese Version of Pepper Spraying Cop?, *Wall Street Journal*, 9 July 2012.

[22] Jessica Levine, Photoshopping Dissent: Circumventing China's Censors With Internet Memes, *The Atlantic*, 4 September 2012.

2.7 Three-North Shelter Forest Program

"Green Great Wall" redirects here. It is not to be confused with the Great Green Wall, a similar anti-desertification effort in Saharan Africa.

The **Three-North Shelter Forest Program** (simplified Chinese: 三北防护林; traditional Chinese: 三北防護林; pinyin: *Sānběi Fánghùlín*), also known as the **Three-North Shelterbelt Program** or the **Green Great Wall**, is a series of human-planted windbreaking forest strips (shelterbelts) in China, designed to hold back the expansion of the Gobi Desert.*[1] It is planned to be completed around 2050,*[2] at which point it will be 2,800 miles (4,500 km) long.

The project's name indicates that it is to be carried out in all three of the northern regions: the North, the Northeast and the Northwest.*[3]

2.7.1 Effects of the Gobi Desert

China has seen 3,600 km² (1,400 sq mi) of grassland overtaken every year by the Gobi Desert.*[4] Each year dust storms blow off as much as 2,000 km² (800 sq mi) of topsoil, and the storms are increasing in severity each year. These storms also have serious agricultural effects for other nearby countries, such as Japan, North Korea, and South Korea.*[5] The Green Wall project was begun in 1978, with the proposed end result of raising northern China's forest cover from 5 to 15 percent *[6] and thereby reducing desertification.

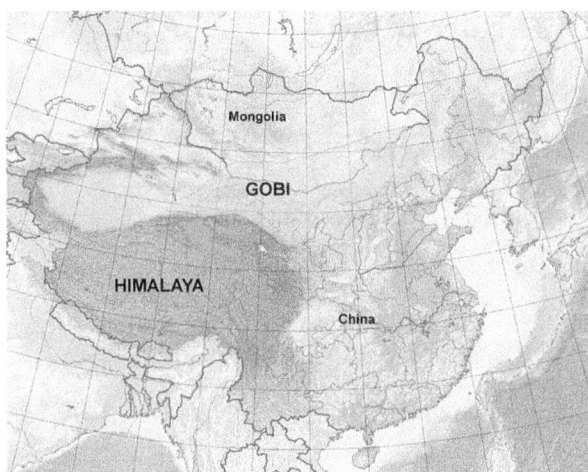

Map of China and the Gobi desert

2.7.2   Methodology and progress

The fourth and most recent phase of the project, started in 2003, has two parts: the use of aerial seeding to cover wide swathes of land where the soil is less arid, and the offering of cash incentives to farmers to plant trees and shrubs in areas that are more arid.[*][7] A $1.2 billion oversight system (including mapping and surveillance databases) is also to be implemented.[*][7] The "wall" will have a belt with sand-tolerant vegetation arranged in checkerboard patterns in order to stabilize the sand dunes. A gravel platform will be next to the vegetation to hold down sand and encourage a soil crust to form.[*][7] The trees should also serve as a windbreak from dust storms.

2.7.3   Measuring success

As of 2009, China's planted forest covered more than 500,000 square kilometers (increasing tree cover from 12% to 18%) – the largest artificial forest in the world.[*][8] Of the 53,000 hectares planted that year, a quarter died.[*][6] In 2008 winter storms destroyed 10% of the new forest stock, causing the World Bank to advise China to focus more on quality rather than quantity in its stock species.[*][8]

2.7.4   Problems

If the trees succeed in taking root, they could soak up large amounts of groundwater, which would be extremely problematic for arid regions like northern China.[*][7] For example, in Minqin, an area in north-western China, studies showed that groundwater levels have dropped by 12–19 metres since the advent of the project.[*][6]

Land erosion and overfarming have halted planting in many areas of the project. China's booming pollution rate has also weakened the soil, causing it to be unusable in many areas.[*][4]

Furthermore, planting blocks of fast-growing trees reduces the biodiversity of forested areas, creating areas that are not suitable to plants and animals normally found in forests. "China plants more trees than the rest of the world combined," says John McKinnon, the head of the EU-China Biodiversity Programme. "But the trouble is they tend to be monoculture plantations. They are not places where birds want to live." The lack of diversity also makes the trees more susceptible to disease, as in 2000, when one billion poplar trees were lost to disease, setting back 20 years of planting efforts.[*][6]

Liu Tuo, head of the desertification control office in the state forestry administration, is of the opinion that there are huge gaps in the country's efforts to reclaim the land that has become desert.[*][9] At present there are around 1.73 million sq kilometers that have become desert in China, of which 530,000 km^2 are treatable. But at the present rate of treating 1,717 km^2 per year, it would take 300 years to reclaim the land that has become desert.[*][10]

2.7.5   Relations to climate change

China's forest scientists argued that monoculture tree plantations are more effective at absorbing the greenhouse gas carbon dioxide than slow-growth forests,[*][8] so while diversity may be lower, the trees purportedly help to offset China's carbon emissions. However, a study released in 2016 finds that wild woodlands are much more effective than monocultural forests in storing carbon dioxide.[*][11]

(See List of countries by carbon dioxide emissions)

2.7.6   Criticism

There are many who do not believe that the Green Wall is an appropriate solution to China's desertification problems. Gao Yuchuan, the Forest Bureau head of Jingbian County, Shanxi, stated that "planting for 10 years is not as good as enclosure for one year," referring to the alternative non-invasive restoration technique that fences off (encloses) a degraded area for two years to allow the land to restore itself.[*][6] Jiang Gaoming, an ecologist from the Chinese Academy of Sciences and proponent of enclosure, says that "planting trees in arid and semi-arid land violates [ecological] principles".[*][6] The worry is that the fragile land cannot support such massive, forced growth. Others worry that China is not doing enough on the social level. In order to succeed, many believe the government should encourage farmers financially to reduce livestock numbers or relocate away from arid areas.[*][7]

2.7.7 See also

- Buffer strip
- Energy-efficient landscaping
- Great Plains Shelterbelt, 1930s-40s, US
- Great Plan for the Transformation of Nature, 1940s-50s, Soviet Union
- Macro-engineering
- Sand fence
- Seawater greenhouse
- Deforestation and climate change

2.7.8 References

[1] "MEDIA REPORTS | China's Great Green Wall". BBC News. 3 March 2001. Retrieved 2012-05-19.

[2] "State Forestry Administration,P.R.China" (in Chinese). English.forestry.gov.cn. Retrieved 2012-05-19.

[3] 李谷城 (Li Kwok-sing) (2006). 中國大陸改革開放新詞語[*A Glossary of New Political Terms of the PRC in the Post-Reform Era*] (in Chinese). HK: Chinese University Press. p. 39. ISBN 978-962-996-258-6.

[4] "The Fall of the Green Wall of China". WorldChanging. 29 December 2003. Retrieved 17 March 2007.

[5] "China's Dust Storms Raise Fears of Impending Catastrophe". National Geographic. 1 June 2001. Retrieved 19 October 2009.

[6] "China's Great Green Wall Proves Hollow". The Epoch Times. 29 July 2009. Retrieved 19 October 2009.

[7] "The Green Wall Of China". Wired. April 2003. Retrieved 19 October 2009.

[8] Watts, Jonathan (11 March 2009). "China's loggers down chainsaws in attempt to regrow forests". London: The Guardian. Retrieved 19 October 2009.

[9] Jonathan Watts (4 January 2011). "China makes gain in battle against desertification but has long fight ahead | Environment". London: The Guardian. Retrieved 2012-05-19.

[10] Patience, Martin (2011-01-04). "BBC News - China official warns of 300-year desertification fight". Bbc.co.uk. Retrieved 2012-05-19.

[11] McGrath, Matt (9 February 2016). "'Wrong type of trees' in Europe increased global warming". BBC News. Retrieved 9 February 2016.

2.7.9 External links

- China's Great Green Wall
- China's forest shelter project dubbed "green Great Wall"
- Grassland ecology to curb sandstorms
- Taming the Yellow Dragon - The Korea Herald
- Verticall gardent in hanoi, vietnam

2.8 Water resources of China

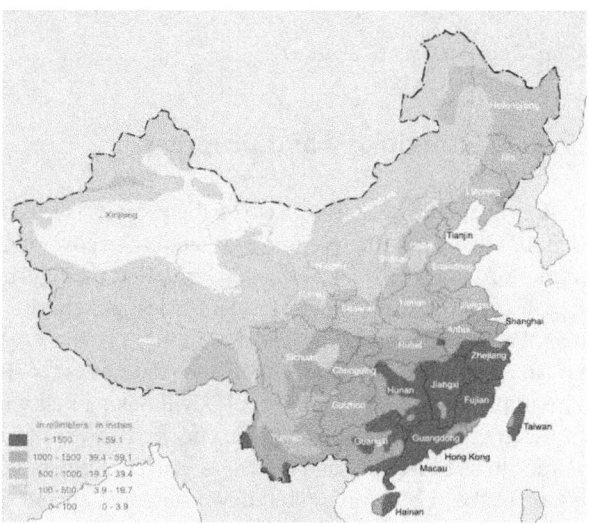

The average annual precipitation in China and Taiwan

The **water resources** of **China** are affected by both severe water quantity shortages and severe water quality pollution. A growing population and rapid economic development as well as lax environmental oversight have increased water demand and pollution. China has responded by measures such as rapidly building out the water infrastructure and increasing regulation as well as exploring a number of further technological solutions.

2.8.1 Water quantity

Supply

China's water resources include 2,711.5 cubic kilometers of mean annual run-off in its rivers and 828.8 cubic kilometers of groundwater recharge. As pumping water draws water from nearby rivers, the total available resource is less than the sum of surface and groundwater, and thus is only 2,821.4 cubic kilometers. 80% of these resources are in the South of China.[*][1]

Demand

Total water withdrawals were estimated at 554 cubic kilometers in 2005, or about 20% of renewable resources. Demand is from the following sectors:

- 65% agriculture

- 23% industry

- 12% domestic

In 2006 626,000 square kilometers were irrigated.[*][1]

Water balance

A farmer's cabbage patch being watered in Linxia County, Gansu

Over-extraction of groundwater and falling water tables are big problems in China, particularly in the north. According to the Ministry of Construction, preliminary statistics show that there are more than 160 areas nationwide where groundwater has been over-exploited with an average annual groundwater depletion of more than 10 billion cubic meters. As a result, more than 60,000 square kilometers of ground surface have sunk with more than 50 cities suffering from serious land subsidence.[*][2] Flooding also still is a major problem.

In a Xinhua article from 2002, Chinese experts warned of future or current water shortages. Water resource usage was expected to peak in 2030 when the population peaks. Areas north of the Yangtze river were particularly affected with 80.9% of Chinese water resources being south of the river. Northern China had used 10,000-year-old aquifers which had resulted in ground cracking and subsidence in some regions.[*][3]

A 2005 article in China Daily stated that out of 514 rivers surveyed in 2000, 60 were dry. Water volume in lakes had decreased by 14%. Many wetlands had decreased in size.[*][4]

Jared Diamond stated in his 2005 book Collapse that, in the past 50 years, exploitation in the form of dams and other irrigation infrastructure have all but halted the Yellow River's natural course, threatening to dry up the entire river valley. The cessation of river flows, or flow stoppages, has surged since the 1980s because of increased water usage and waste. In 1997, the lower Yellow River did not flow 230 days out of the year, an increase of over 2000% since 1988. Increased erosion and sedimentation, especially on the Loess Plateau, has made the river much less navigable by ship.[*][5]

For the 2008 Summer Olympics, China diverted water from Hebei and Shanxi provinces, areas already beset by drought and dramatic water shortages, to Beijing.[*][6] In July 2008, the head of the Beijing Water Authority Bi Xiaogang denied that the Olympics would increase water consumption by a large amount. However, previously he and other local officials said that Beijing would divert up to 400 million cubic meters of water from Hebei for the Games with water-diversion facilities and pipes being built to pump water from four reservoirs in Hebei.[*][7] Around Baoding city alone, a mostly rural area, 31,000 residents lost land and their homes due to a water transfer project; many more have been displaced throughout Hebei.[*][8][*][9] According to an August 24, 2008 report by the UK's *Times,* much of the infrastructure intended for the water diversion scheme was left half-constructed or unused when Beijing officials realized that water demand estimates had been far too high. The number of tourists attending the Beijing games was lower than expected, and many migrant workers, ethnic minorities, and political dissidents had left the city as a result of intimidation or official requests. Nevertheless, the Hebei area had already been sucked dry to fill a number of large reservoirs, leading to drought and agricultural losses.[*][10]

Water transfers

Large-scale water transfers have long been advocated by Chinese planners as a solution to the country's water woes. The South-North Water Transfer Project is being developed primarily to divert water from the Yangtze River to the Yellow River and Beijing.

The development or diversion of major transboundary rivers originating in China, such as the Brahmaputra River and the Mekong River, could be a source on tension with China's neighbors. For example, after building two dams upstream, China is building at least three more on the Mekong, inflaming passions in Vietnam, Laos, Cambodia and Thailand. In a book titled "Tibet's Waters Will Save China" a group of Chinese ex-officials have championed the northward rerouting of the waters of the Brahmaputra as an important lifeline for China in a future phase of South-North Water Transfer Project. Such a diversion could fuel tension with India and Bangladesh, if no prior agreement would be reached on sharing the river's water.[11]

On a smaller scale, some of the waters of the Irtysh River, which would otherwise flow into Kazakhstan, Russia, and the Arctic Ocean, have been diverted into the arid areas of north-central Xinjiang via the Irtysh–Karamay–Ürümqi Canal.

Sea water desalination

Due to the water problems, as well as for future exports, China is building up its desalination technological abilities and plans to create an indigenous industry. Some cities have introduced extensive water conservation and recycling programs and technologies.[12]

2.8.2 Water quality

The quality of groundwater or surface water is a major problem in China, be it because of man-made water pollution or natural contamination.

Pollution and Water Shortage

See also: Yellow River § Pollution and Yangtze River § Degradation of the river
Deterioration of drinking water quality continues to be a

Industrial and domestic development along the Yellow River at Liujiaxia Dam

major problem in China. Continuous emissions from manufacturing is the largest contributor to lowered drinking quality across the People's Republic,[13] but introduction

An almost-dry river near Beijing, China. July 2007

of poorly treated sewage, industrial spills, and extensive use of agricultural fertilizers and pesticides have proven to be major contributors as well. Furthermore, these water quality issues couple with seasonal scarcity of water to spark endemic water shortages, which frequently affect millions of people to some extent.[14]

According to China's State Environmental Protection Administration (SEPA) in 2006, 60% of the country's rivers suffer from pollution to such an extent that they cannot be safely used as drinking water sources.[15] According to the 2008 State of the Environment Report by the Ministry of Environmental Protection, the successor agency of SEPA, pollution of specific rivers is as follows:

- The Pearl River and the Yangtze River had "good water quality";

- The Songhua River was "slightly polluted" (it was "moderately polluted" in 2006);

- The Liaohe River, the Huaihe River, and the Yellow River were "moderately polluted" (another translation says they "had poor water quality"); and

- the Haihe River which flows through Beijing and Tianjin was "badly polluted".[16]

A 2006 article by the Chinese Embassy in the UK stated that approximately 300 million nationwide have no access to clean water. Almost 90% of underground water in cities are affected by pollution and as well as 70% of China's rivers and lakes.[17]

A 2007 article in China Daily stated that large scale use of pesticides and fertilizers from agriculture also contribute to water pollution.[18]

A 2008 report about the Yellow River argued that severe pollution caused by factory discharges and sewage from fast-expanding cities has made one-third of the river unusable even for agricultural or industrial use. The report, based on data taken last year[which year?], covered more than 8,384 miles of the river, one of the longest waterways in the world, and its tributaries. The Yellow River Conservancy Committee, which surveyed more than 8,384 miles of the river in 2007, said 33.8% of the river system registered worse than level five. According to criteria used by the UN Environment Program, level five is unfit for drinking, aquaculture, industrial use and even agriculture. The report said waste and sewage discharged into the system last year totaled 4.29bn tonnes. Industry and manufacturing provided 70% of the discharge into the river, with households accounting for 23% and just over 6% coming from other sources.[13]

There have been a high number of river pollution incidents in recent years in China, such as drinking water source pollution by algae in the Tai Lake, Wuxi in May 2007. There was a "bloom of blue-green algae that gave off a rotten smell" shutting off the main source of drinking water supply to 5.8 million people. By September 2007, the city had closed or given notice to close more than 1,340 polluting factories. The city ordered the rest to clean up by June or be permanently shut down. The closing of the factories resulted in a 15% reduction of local GDP.[19] The severe pollution had been known for many years, but factories had been allowed to continue to operate until the crisis erupted.

The 2005 Jilin chemical plant explosions in Jilin City caused a large discharge of nitrobenzene into the Songhua River. Levels of the carcinogen were so high that the entire water supply to Harbin city (pop 3.8M) was cut off for five days between November 21, 2005 and November 26, 2005, though it was only on November 23 that officials admitted that a severe pollution incident was the reason for the cutoff.[20]

Chinese environmental activist and journalist Ma Jun warned in 2006 that China is facing a water crisis that includes water shortages, water pollution and a deterioration in water quality. Ma argued that 400 out of 600 cities in China are facing water shortages to varying degrees, including 30 out of the 32 largest cities. Furthermore, Ma argued, discharges of waste water have increased continually over the years 2001-2006, and that that 300 million peasants' drinking water is not safe. He warned: "In the north, due to the drying up of the surface water, the underground water has been over-extracted. The water shortage in the north could have drastic affects because almost half of China's population lives on only 15 percent of its water. The situation is not sustainable. Though the south has abundant water, there is a lack of clean water due to serious water pollution. Even water-abundant deltas like the Yangtze and the Pearl River suffer from water shortages." [21][22]

According to an article in the Guardian, in 2005, deputy minister Qiu Baoxing stated that more than 100 out of the 660 cities had extreme water shortages. Pan Yue, deputy director of the state environment protection agency, warned that economic growth was unsustainable due to the water problems. In 2004 the World Bank warned that the scarcity of the resource would lead to "a fight between rural interests, urban interests and industrial interests on who gets water in China." In April 2005 there were dozens of injuries in Dongyang city, Zhejiang Province, due to clashes over the nearby chemical factories of the Juxi Industrial Park accused of water pollution that harmed crops and led to deformed babies being born. According to the article, a quarter of the population lacked clean drinking water and less than a third of the waste was treated. China is expected to face worsening water shortages until 2030 when the population peaks.[14]

According to a 2007 report by the World Bank, the pollution scandals demonstrate that, if not immediately and effectively controlled, pollution releases can spread across the boundaries of administrative jurisdictions, causing "environmental and economic damage as well as public concern and the potential for social unease" . Once an accident has occurred, the impact on the environment and human health becomes more difficult and more costly to control. Therefore, the report recommends prevention of pollution by strict enforcement of appropriate policies and regulations.[23]

Natural contamination

Large portions of China's aquifers suffer from arsenic contamination of groundwater. Arsenic poisoning occurs after long-term exposure to contaminated groundwater through drinking. The phenomenon was first detected in China in the 1950s. As water demand grows, wells are being drilled deeper and now frequently tap into arsenic-rich aquifers. As a consequence, arsenic poisoning is rising. To date there have been more than 30,000 cases reported with about 25 million people exposed to dangerously high levels in their drinking water.[24]

According to the WHO over 26 million people in China suffer from dental fluorosis (weakening of teeth) due to elevated fluoride in their drinking water. In addition, over 1 million cases of skeletal fluorosis (weakening of bones) are thought to be attributable to drinking water.[25] High levels of fluoride occur in groundwater and defluoridation is in many cases unaffordable.

Pollution incidents

The Hubei Shuanghuan Science and Technology Stock Co poisoned at least 100 tonnes (220,000 lb) fish in a river in central Hubei province in September 2013 with discharged ammonia into the Fuhe river.[*][26]

2.8.3 Conservation and Sanitation

Main article: Water supply and sanitation in the People's Republic of China

Water supply and sanitation in the People's Republic of China is undergoing a massive transition, while facing numerous challenges - such as rapid urbanization and a widening economic gap between urban and rural areas.[*][27]

The World Bank in a 2007 report stated that between 1990 and 2005 there have been major financial investments in water infrastructure. While urban water supply coverage increased from 50% to 90%, there are still seasonal water shortages in many cities. Water usage by the growing population has increased but it has decreased by industry causing a stabilization of the overall water usage level. Sewage treatment of urban wastewater more than tripled from 15% to 52%. Installed wastewater treatment capacity grew much more quickly due to an increasing absolute amount of wastewater. Absolute release of municipal pollutants has decreased slightly since 2000.[*][28]

According to a 2007 article, the SEPA stated that the water quality in the central drinking water sources for major cities was "mainly good".[*][15]

2.8.4 Management

The responsibility for dealing with water is split between several agencies within the government. Water pollution is the responsibility of the environmental authorities, but the water supply itself is managed by the Ministry of Water Resources. Sewage treatment is managed by the Ministry of Construction, but groundwater management falls within the realm of the Ministry of Land and Resources. China grades its water quality in six levels, from Grade I to Grade VI, with Grade VI being the most polluted.[*][29]

In 2007 Ma Xiancong, a researcher at the Chinese Academy of Social Sciences Institute of Law, identified the following areas where the government failed to act, or tacitly consented, approved or actively took part and so created a worse situation: land appropriation, pollution, excessive mining and the failure to carry out environmental impact assessments. An example of this emerged in 2006, when the State Environmental Protection Administration revealed over a dozen hydroelectric projects that had broken the Environmental Impact Assessment Law.[*][30]

In 2005 experts warned that China must use Integrated Water Resources Management in order to achieve sustainable development.[*][4]

2.8.5 See also

- Water supply and sanitation in China
- Water supply and sanitation in Hong Kong
- Ministry of Environmental Protection of the People's Republic of China
- Geography of China
- Agriculture in China
- Environment of China

2.8.6 References

[1] FAO Aquastat:China Profile, Version 2010

[2] China Development Gateway: Ensuring the Safety of Urban Water Supply, Facilitating the Frugal and Appropriate Consumption of Urban Water, Ministry of Construction, August 22, 2006 MOC

[3] China Warned of Water Crisis by 2030, "china.org.cn", June 6, 2002

[4] Experts warn of water crisis, *China Daily*, May 20, 2005

[5] Diamond, Jared: "Collapse," pp.364-5. Penguin Books, 2005

[6] Chris Buckley (2008). "Beijing Olympic water scheme drains parched farmers." Reuters, January 22, 2008.

[7] Shi Jiangtao (2008). "Official Denies Plan to Divert Water from Parched Provinces." South China Morning Post, July 26, 2008.

[8] Xinhua (2008). "China refills lake." June 20, 2008

[9] Xinhua (2007). "Hebei Reservoirs." November 26, 2007.

[10] Michael Sheridan (2008). "Millions forfeit water to Olympic Games." *Times,* August 24, 2008.

[11] China Aims for Bigger Share of South Asia's Water Lifeline, by Brahma Chellaney, Japan Times, June 26, 2007

[12] MICHAEL WINES, China Takes a Loss to Get Ahead in the Business of Fresh Water, October 25, 2011, http://www.nytimes.com/2011/10/26/world/asia/china-takes-loss-to-get-ahead-in-desalination-industry.html?pagewanted=all

[13] Tania Branigan (25 November 2008). "One-third of China's Yellow river 'unfit for drinking or agriculture' Factory waste and sewage from growing cities has severely polluted major waterway, according to Chinese research". London: guardian.co.uk. Retrieved 2009-03-14.

[14] 100 Chinese cities face water crisis, *The Guardian*, June 8, 2005

[15] "China pays water price for progress", Water 21, Magazine of the International Water Association, August 2007, p. 6

[16] Ministry of Environmental Protection:The State of the Environment of China in 2008, June 5, 2009

[17] Miao Hong (2006). "China battles pollution amid full-speed economic growth." Chinese Embassy (UK), September 29, 2006.

[18] "Pollution makes cancer th54uy6u56u56u356u56u56u56e top killer". Xie Chuanjiao (China Daily). 2007-05-21.

[19] Washington Post:In China, a Green Awakening City Clamps Down on the Polluting Factories That Built Its Econonomy, October 6, 2007, p. A1, accessed on October 14, 2007

[20] "China city water supply to resume". BBC. 2005-11-27.

[21] Tackling China's water crisis

[22] Larmer, Brook. (May 2008). Bitter Waters. National Geographic Retrieved on 20 January 2009.

[23] World Bank (2007):Water Pollution Emergencies in China - Prevention and Response accessed on September 4, 2007

[24] UNICEF:China:Child's environment and sanitation, accessed on December 24, 2009

[25] WHO:Facts and figures: Water, sanitation and hygiene links to health, accessed on December 24, 2009

[26] China chemical spill kills thousands of fish bbc 4 September 2013

[27] BBC News. China to clean up polluted lake. 27 October 2007.

[28] World Bank:Stepping up - Improving the performance of China's urban water utilities, by Greg Browder et al., 2007

[29] Ma, Xiangcong (2007-02-21). "China's environmental governance". chinadialogue.

[30] Chinadialogue: "China's environmental governance", Ma Xiangcong, February 21, 2007; retrieved on 25 October 2011

2.8.7 External links

- Peter Gleick:*China and Water*, Chapter 5 of The World's Water 2008-2009, Pacific Institute, 2009.

- Jian Xie, with Andres Liebenthal, Jeremy J. Warford, John A. Dixon, Manchuan Wang, Shiji Gao, Shuilin Wang, Yong Jiang, and Zhong Ma:*Addressing China's Water Scarcity. Recommendations for Selected Water Resource Management Issues*, The World Bank, 2009

- Environment & Energy News at China Development Gateway

- World Bank: "Agenda for Water Sector Strategy for North China", April 2001

- "China to step up water resource protection". People's Daily. 22 March 2005.

- "Chinese cities face toxic spills: Explosions at chemical plants leave millions without clean water" Subscription required *Nature*, November 25, 2005

- Chinadialogue:Lee Seungho: "Wet politics in China" 中国与世界，环境危机大家谈 September 25, 2006

- Chinadialogue:Ma Xiangcong: "China's environmental governance", February 21, 2007

- China water pollution - Greenpeace China

- Ministry of Environmental Protection

- chinadialogue 中国与世界,环境危机大家谈 - bilingual news and in-depth articles on China's environmental crisis

- China's Water Management An interview of Joël Cicéron, director of Veolia Water in Taiwan.

- River Pollution in China Photo essay on water pollution in the Huai River Basin

- Worsening Water Shortages Threaten China's Food Security

- Water issues in China, from PBS site

- CNN.com audio slideshow on water pollution in China

- *Shifting Nature*. Produced by BBC Two.

Chapter 3

Climate change in China

3.1 China Beijing Environmental Exchange

China Beijing Environmental Exchange (CBEEX) is a corporate domestic and international environmental equity public trading platform initiated by the China Beijing Equity Exchange (CBEX) and authorized by the Beijing municipal government.

In June 2009, the CBEEX signed a deal with BlueNext to build a platform for carbon credit trade. BlueNext is Europe's largest carbon credit exchange and is owned jointly by NYSE Euronext and French state-owned bank Caisse des Dépôts.[1][2][3][4][5]

3.1.1 Aims

3.1.2 See also

- Tianjin Climate Exchange, set up by the Chicago Climate Exchange and China National Petroleum Corp

3.1.3 References

[1] Xinhua

[2] Financial Times

[3] Reuters

[4] Wall Street Journal

[5] Exchange News Direct

3.1.4 External links

- Official website

3.2 Climate change in China

The position of the Chinese government on climate change is contentious. China has ratified the Kyoto Protocol, but as a non-Annex I country which is not required to limit greenhouse gas emissions under terms of the agreement. In particular since 2007 the Chinese government has changed its attitude towards climate change policy and has become one of the major drivers of low-carbon technology developments.[1]

In 2002, on the basis of an analysis of fossil fuel consumption (including especially the coal power plants[2]) and cement production data, that China surpassed the United States as the world's largest emitter of carbon dioxide, putting out 7,000 million tonnes, in comparison with America's 5,800 million.[3]

According to data from the US Energy Information Administration China was the top emitter by fossil fuels CO2 in 2009 China: 7,710 million tonnes (mt) (25.4%) ahead of US: 5,420 mt (17.8%), India: 5.3%, Russia: 5.2% and Japan: 3.6%.[4]

China was also the top emitter of all greenhouse gas emissions including building and deforestation in 2005: China: 7,220 mt (16.4%), US: 6,930 mt (15.7%), 3. Brazil 6.5%, 4. Indonesia: 4.6%, 5. Russia 4.6%, 6. India 4.2%, 7. Japan 3.1%, 8. Germany 2.3%, 9. Canada 1.8%, and 10. Mexico 1.6%.[4]

In the cumulative emissions between 1850 and 2007 the top emitters were: 1. US 28.8% 2. China: 9.0%, 3. Russia 8.0%, 4. Germany 6.9%, 5. UK 5.8%, 6. Japan 3.9%, 7. France 2.8%, 8. India 2.4%, 9.bvanan Canada 2.2% and 10. Ukraine 2.2%.[4]

According to BBC News, in September 2014, China surpassed the European Union's per capita carbon emissions for the first time in history. China's per capita carbon emissions now stand at 7.2 t/capita. [5] China's carbon emissions have increased rapidly since its economic boom in the early 2000s. Since then, their per capita carbon emissions have increased by more than 2.5 times. [6]

3.2.1   Total emissions

According to a statement made in The Economist in 2013, China has emitted more climate change gases from energy production than America since 2006 and by 2014-2015 China will emit twice America's total. At the present rate of development, cumulative Chinese emissions from energy production between 1990 and 2050 will equal those generated by the whole world from the beginning of the industrial revolution to 1970. About a quarter of China's carbon emissions are produced in the manufacture of goods for export. [7]

3.2.2   Coal

See also: Coal power in China

China is the largest consumer of coal in the world.

In 2009, China produced 18,449 TWh of the world's total 39,340 TWh. [8]

China is now adding sulfur dioxide reducing technology to its power plants. It has been argued that the release of sulfur dioxide from burning coal has slowed global warming but has caused 4,698.3 deaths in the past decade. [9]

3.2.3   Effects of climate change

China can suffer some of the effects of global warming, including sea level rise, glacier retreat and air pollution.

The implications of climate change impose serious setbacks on global health and will hinder the economic development of various regions worldwide impacting countries on more than just the basic environmental scale. As in the case of China, we will see the effects on a social and economic level.

China's first National Assessment of Global Climate Change, released recently by the Ministry of Science and Technology (MOST), states that China already suffers from the environmental impacts of climate change: increase of surface and ocean temperature, rise of sea level. [10] Qin Dahe,former head of China's Meteorological Administration, has said that the temperatures in the Tibetan Plateau of China are rising four times faster than anywhere else. [11] Rising sea level is an alarming trend because China has a very long and densely populated coastline, with some of the most economically developed cities such as Shanghai, Tianjin, and Guangzhou situated there. Chinese research has estimated that a one-meter rise in sea level would inun-

date 92,000 square kilometres of China's coast, thereby displacing 67 million. [12]

There has also been an increased occurrence of climate-related disasters such as drought and flood and the amplitude is growing. They have grave consequences for productivity when they occur, and also create serious repercussions for natural environment and infrastructure. This threatens the lives of billions and aggravates poverty.

Furthermore, climate change will worsen the uneven distribution of water resources in China. Outstanding rises in temperature would exacerbate evapo-transpiration intensifying the risk of water shortage for agricultural production in the North. While because of the southern region's over abundance in rainfall, most of its water is lost due to flooding. As the Chinese government faces challenges managing its expanding population, an increased demand for water to support the nation's economic activity and people will burden the government. In essence, a water shortage is indeed a large concern for the country. [12]

Lastly, climate change could endanger human health by increasing outbreaks of disease and their transmission. After floods, for example, infectious diseases such as diarrhea, cholera are all far more prevalent. These effects would exacerbate the degradation of the ecologically fragile areas in which poor communities are concentrated pushing thousands back into poverty. [13]

Agriculture

1 °C of regional mean warming is estimated to reduce wheat yield 3 to 10 percent in China. Grain crops mature earlier at higher temperatures, reducing the critical growth period and leading to lower yields (You et al. 2009). [14]

Diseases

Some regions in China will be exposed to a 50 percent higher malaria transmission probability rate (Béguin et al., 2011). [14]

IPCC

According to IPCC (2007) from 1900 to 2005 precipitation has declined in parts of southern Asia. By the 2050s freshwater availability including large river basins is projected to decrease in Asian regions. Coastal areas, specially the delta areas in Asia are projected to have increased flooding risk. Floods and droughts are expected to increase health concerns: diseases and mortality. [15]

3.2.4 Debate over China's economic responsibilities for climate change mitigation

Main article: Debate over China's economic responsibilities for climate change mitigation

Both internationally and within the People's Republic of China, there has been an ongoing debate over China's economic responsibilities for climate change mitigation.

3.2.5 Climate change mitigation measures

The People's Republic of China is an active participant in the climate change talks and other multilateral environmental negotiations, and claims to take environmental challenges seriously but is pushing for the developed world to help developing countries to a greater extent. It is a signatory to the Kyoto Protocol, although China is not required to reduce its carbon emissions under the terms of the present agreement.

China issued is first Climate Change Program in 2007, in response to its surpassing of the United States as the largest emitter of carbon dioxide emissions in the world.[16] The Chinese national carbon trading scheme was later announced in November 2008 by the national government to enforce a compulsory carbon emission trading scheme across the country's provinces as part of its strategy to create a "low carbon civilisation".[17] The scheme would allow provinces to earn money by investing in carbon capture systems in those regions that fail to invest in the technology.[18]

In 2004, Premier Wen Jiabao promised to use an "iron hand" to make China more energy efficient. China has surpassed the rest of the world as the biggest investor in wind turbines and other renewable energy technology. And it has dictated tough new energy standards for lighting and gas kilometrage for cars.[19] With $34.6 billion invested in clean technology in 2009, China is the world's leading investor in renewable energy technologies.[20][21] China produces more wind turbines and solar panels each year than any other country.[22]

Coal is predicted to remain the most important power source in the near future but China has been seen as the world leader in clean coal technology.[23][24][25]

Nuclear power is planned to be rapidly expanded. By mid-century fast neutron reactors are seen as the main nuclear power technology which allows much more efficient use of fuel resources.[26]

China should push electric cars to curb its dependence on imported petroleum (oil) and foreign automobile technology, although they offer smaller cuts in carbon emissions than alternatives like hybrid electric vehicles, consulting firm McKinsey & Co says.[27]

A 2011 Lawrence Berkeley National Laboratory report predicted that Chinese CO_2 emissions will peak around 2030. This because in many areas such as infrastructure, housing, commercial building, appliances per household, fertilizers, and cement production a maximum intensity will be reached and replacement will take the place of new demand. The 2030 emissions peak also became China's pledge at the Paris COP21 summit. Carbon emission intensity may decrease as policies become strengthened and more effectively implemented, including by more effective financial incentives, and as less carbon intensive energy supplies are deployed. In a "baseline" computer model CO_2 emissions were predicted to peak in 2033; in an "Accelerated Improvement Scenario" they were predicted to peak in 2027.[28]

3.2.6 See also

- Chinese national carbon trading scheme
- China Carbon Forum
- China Beijing Environmental Exchange
- Coal power in China
- Deforestation and climate change
- Environment of China
- Politics of global warming
- Renewable energy in China
- Renewable energy commercialization
- Solar power in China
- Scientific Development Concept
- Tianjin Climate Exchange

3.2.7 References

[1] Gippner, Olivia (2014) Framing it right: China-EU relations and patterns of interaction on climate change Chinese Journal of Urban and Environmental Studies, 2 (1). ISSN 2345-7481

[2] Coal power plants in China Map +oil use. Platts.com (1999-02-22). Retrieved on 22 September 2002.

[3] "China now no. 1 in CO_2 emissions; USA in second position". Netherlands Environmental Assessment Agency. 19 June 2007.

[4] Which nations are most responsible for climate change? Guardian 21 April 2011

[5] , BBC News 21 September 2014

[6] , 23 September 2014,

[7] The east is grey The Economist 10 August 2013

[8] name=Sverigetab49>Energy in Sweden 2010, Facts and figures, The Swedish Energy Agency, Table 52: Global supply of coal 1990–2009 (TWh)

[9] David Biello, Stratospheric Pollution Helps Slow Global Warming, 22 July 2011, Scientific American, http://www.scientificamerican.com/article.cfm?id= stratospheric-pollution-helps-slow-global-warming

[10] http://www.die-gdi.de/CMS-Homepage/openwebcms3_ e.nsf/(ynDK_contentByKey)/ENTR-7BDE2T? OpenDocument&nav=expand:Research% 20and%20Consulting__xunadd_text_character: nN{\textbackslash}{\}{ }Projects;active:Research% 20and%20Consulting__xunadd_text_character: nN{\textbackslash}{\}{ }Projects__xunadd_text_ character:nN{\textbackslash}{\}{ }ENTR-7BDE2T

[11] The Indus Equation Report, Strategic Foresight Group

[12] http://stonybrook.digication.com/egimenez/Case_Study_ Impact_of_Climate_Change_on_China

[13] http://www.oxfam.org.hk/en/climatepoverty.aspx

[14] Why a 4 degree centrigrade warmer world must be avoided November 2012 World Bank

[15] IPCC Working group III fourth assessment report, Summary for Policymakers 2007

[16] Andrews-Speed, Philip (November 2014). "China's Energy Policymaking Processes and Their Consequences". *The National Bureau of Asian Research Energy Security Report*. Retrieved 5 December 2014.

[17] Climate Ark:China outlines plans for domestic carbon trading

[18] businessgreen.com: China outlines plans for domestic carbon trading

[19] Bradsher, Keith (4 July 2010). "China Fears Warming Effects of Consumer Wants". *The New York Times*.

[20] China Leads Major Countries With $34.6 Billion Invested in Clean Technology

[21] China steams ahead on clean energy

[22] Bradsher, Keith, 30 January 2010, China leads global race to make clean energy, New York Times

[23] JAMES FALLOWS, Dirty Coal, Clean Future, DECEMBER 2010 ATLANTIC MAGAZINE, http://www.theatlantic.com/magazine/archive/2010/ 12/dirty-coal-clean-future/8307/1/

[24] China's coal reserves 'will make it new Middle East', says energy chief, Leo Hickman, Tuesday 8 March 2011, The Guardian, http://www.guardian.co.uk/environment/ 2011/mar/08/china-coal-new-middle-east

[25] KEITH BRADSHER, China Outpaces U.S. in Cleaner Coal-Fired Plants, 10 May 2009, The New York Times, http://www.nytimes.com/2009/05/11/world/asia/11coal. html

[26] Nuclear Power in China, Updated March 2012, World Nuclear Association, http://www.world-nuclear.org/info/ inf63.html

[27] LexisNexis® Publisher

[28] ChinaFAQs: China's Energy and Carbon Emissions Outlook to 2050, ChinaFAQs on 12 May 2011, http://www.chinafaqs.org/library/ chinafaqs-chinas-energy-and-carbon-emissions-outlook-2050

3.2.8 External links

- China Climate Change Information Network

- Greenpeace China's China climate change resources

- China-US Energy Efficiency Alliance: China Climate Change News & Resources

3.3 Coal in China

Entrance to a small coal mine in China.

China is the largest producer and consumer of **coal** in the world and is the largest user of coal-derived electricity, generating an estimated 73% of domestic electricity production in 2014 from coal.[1][2]

Both coal production and consumption peaked in 2013 and has dropped continuously, falling a further 3.7% in the first 11 months of 2015 compared to the same period the year before.[3][4]

A coal shipment underway in China.

An operating power plant in China.

The central government has curbed construction of new coal fired plants with national regulators ordering in April 2016 a halt to construction in 13 provinces and delays for already approved projects in a further 15 provinces.*[5] This is in line with a moratorium issued by the National Energy Agency in 2015 banning new coal mines in China for a period of three years and closure of thousands of small coal mines.*[6]

3.3.1 Resource flow

Coal reserves

As of the end of 2014, China had 62 billion tons of anthracite and 52 billion tons of lignite quality coal. China ranks third in the world in terms of total coal reserves behind the United States and Russia.*[7] Most reserves are located in the north and north-west of the country, which poses a large logistical problem for supplying electricity to

Coal reserves in BTUs as of 2009

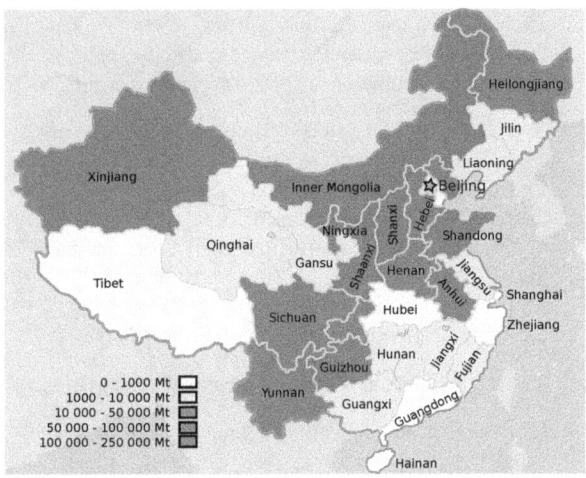

Coal resources in China (2001)

the more heavily populated coastal areas.*[8] At current levels of production, China has 30 years worth of reserves.*[9] However, others suggest that China has enough coal to sustain its economic growth for a century or more.*[10]

Coal production

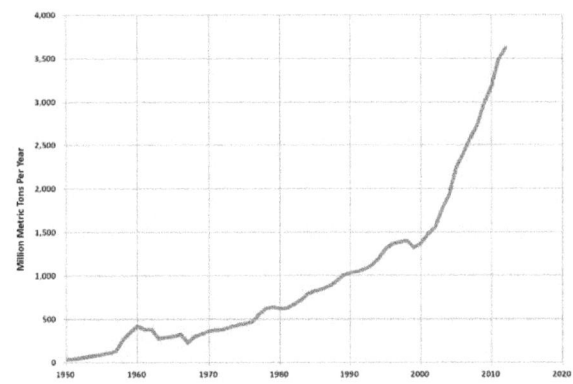

Coal production in China, 1950-2012

China is the largest coal producer in the world,*[11] but as

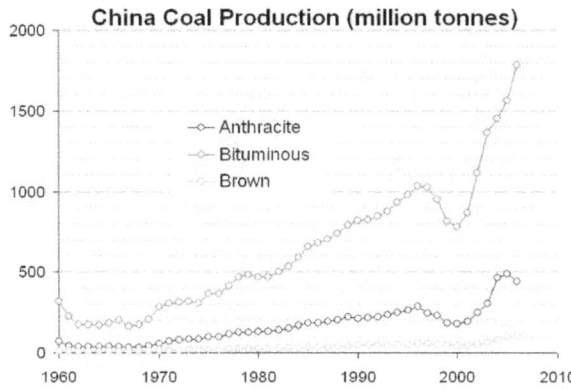

Production of coal within China by type.

For reference: GDP of the PRC. Coal production and usage demonstrates a hypersensitivity to economic changes.

of 2015 falling coal prices resulted in layoffs at coal mines in the northeast.[12]

Coal is the major source of energy in China. In 2011 the Chinese coal production was equivalent to $3{,}576$ Mt $\times 0.522$ toe/t $\times 11.630$ MWh/toe $= 21{,}709$ TWh. Assuming the same caloric value for the imported coal the net coal energy available would be evaluated as 22,784 TWh. Assuming imported coal equal to domestic one, available coal (IEA) was about 17,000 TWh in 2008 and 22,800 TWh in 2011, with increase of 5,800 TWh in three years. Total renewable energy in China was 3,027 TWh in 2008 and 2,761 TWh in 2005, with increase of 266 TWh in three years. Same period from 2005 to 2008 annual coal use increased 3,341 TWh.

Energy demand in China continues to increase, with electric demand roughly doubled to 2013,[7] The demand for coal in China had increased so fast, demand had exceed production due to factors such as a government crackdown on mines that are unsafe, polluting, or wasteful. Some were shut down for the 2008 Summer Olympics.[14]

On July 6, 2008 in central and northern China, 2.5% of the nation's coal plants (58 units or 14,020 MW of capacity) had to shut down due to coal shortages. This forced local governments to limit electricity consumption and issue blackout warnings. The shortage is somewhat attributed to the closing of small dangerous coal mines.[15]

In 2011, seven Chinese coal mining companies produced 100 million metric tonnes of coal or more. These companies were Shenhua Group, ChinaCoal, Shaanxi Coal and Chemical Industry, Shanxi Coking Coal Group, Datong Coal Mine Group, Jizhong Energy, and Shandong Energy.[16] The largest metallurgical coal producer was Shanxi Coking Coal Group.[17]

In 2015, official statistics revealed that previous statistics had been systematically underestimated by 17%, corresponding to the entire CO_2 emissions of Germany.[18]

A coal mine near Hailar.

Inner Mongolia   China's largest open-pit coal mine is located in Haerwusu in the Inner Mongolia Autonomous Region. It started production on 20 October 2008, and is operated by Shenhua Group. Its estimated coal output was forecast at 7 million tonnes in the fourth quarter of 2008. With a designed annual capacity of 20 million tonnes of crude coal, it will operate for approximately 79 years. Its coal reserves total about 1.73 billion tonnes. It is rich in low-sulfur steam coal.[19] Mines in Inner Mongolia are rapidly expanding production, with 637 million tons produced in 2009. Transport of coal from this region to seaports on China's coast has overloaded highways such as China National Highway 110 resulting in chronic traffic jams and delays.[20]

3.3.2   Coal consumption

China's coal consumption in 2010 was 3.2 billion metric tonnes per annum. The National Development and Reform Commission, which determines the energy policy of China, aims to keep China's coal consumption below 3.8 billion metric tonnes per annum.

With investment in the coal industry rising at an annual rate of 50 percent in recent years, China will retain its current position as the leading global consumer of coal, even as it endeavors to diversify.

During the first three quarters of 2009 China's coal consumption increased 9% from 2008 to 2.01 billion metric tons.[21]

The consumption of coal is largely in power production, aside from this, there is a lot of industry and manufacturing use along with a comparatively very small amount of domestic use.

Coal for domestic use being transported by use of a tricycle.

Electricity generation

Main article: Electricity sector in China

Coal power is distributed by the State Power Grid Corporation.

China's installed coal-based electrical capacity was 907 GW, or 77% of the total electrical capacity, in 2014.[24] [25] The dominant technology in the country is coal pulverization in lieu of the more advanced and preferred coal gasification. China's move to a more open economy in the 1990s is cited as a reason for this, where the more immediately lucrative pulverization technology was favored by businesses. There are plans in place for an Integrated Gasification Combined Cycle (IGCC) type plant by 2010.[26] Furthermore, less than 15% of plants have desulphurization systems.[27]

Industrial use

China's energy consumption is mostly driven by the industry sector, the majority of which comes from coal consumption.[28] One of the principal users is the steel industry in China.

Domestic use

In cities the domestic burning of coal is no longer permitted. In rural areas coal is still permitted to be used by Chinese households, commonly burned raw in unvented stoves. This fills houses with high levels of toxic metals leading to bad Indoor Air Quality (IAQ). In addition, people eat food cooked over coal fires which contains toxic substances. Toxic substances from coal burning include arsenic, fluorine, polycyclic aromatic hydrocarbons, and mercury. Health issues are caused which include severe arsenic poisoning, skeletal fluorosis (over 10 million people afflicted in China), esophageal and lung cancers, and selenium poisoning.[29]

In 2007 the use of coal and biomass (collectively referred to as solid fuels) for domestic purposes was nearly ubiquitous in rural households but declining in urban homes. At that time, estimates put the number of premature deaths due to indoor air pollution at 420,000 per year, which is even higher than due to outdoor air pollution, estimated at around 300,000 deaths per year. The specific mechanisms for death cited have been respiratory illnesses, lung cancer, Chronic Obstructive Pulmonary Disease (COPD), weakening of the immune system, and reduction in lung function. Measured pollution levels in homes using solid fuels generally exceeded China's IAQ air quality standards. Technologies exist to improve indoor air quality, notably the installation of a chimney and modernized bioenergy but need more support to make a larger difference.[30]

International trade

China became a net importer of coal in 2008.[31] In 2006, its exports exceeded imports by 25.1 million tons, but only by 2 million tons in 2007. This is significantly lower than the 90 million ton net exports in 2001.

Vietnam is the largest supplier of coal to China at 24.6 million tonnes for 2007.[31] Australia exported 4.52 million tonnes in 2007.[31]

In the first quarter of 2015, China's coal imports fell 42% from the previous year, due to a slower economy and tougher air pollution standards.[32]

3.3.3 Carbon footprint

In 2014 the carbon emissions from China made up about 28.8% of the world total, 10.4 billion tons.CO_2 emissions [*][33]

It is believed that a continued increase in coal power in China may undermine international initiatives to decrease carbon emissions such as the Kyoto Protocol, which called for a decrease of 483 million tons by 2012. In the same time frame, it is expected that coal plants in China will have increased CO_2 emissions by 1,926 million tons —over 4 times the proposed reduction.[*][34]

Efforts to reduce emissions

Air pollution has gotten so bad that a study by the World Bank found that air pollution kills 750,000 people every year in China.[*][36] Issued in response to record-high levels of air pollution in 2012 and 2013, the State Council's September 2013 Action Plan for the Prevention and Control of Air Pollution reiterated the need to reduce coal's share in China's energy mix to 65% by 2017.[*][37] Amidst growing public concern, social unrest incidents are growing around the country. For example, in December 2011 the government suspended plans to expand a coal-fired power plant in the city of Haimen after 30,000 local residents staged a violent protest against it, on the grounds that "the coal-fired power plant was behind a rise in the number of local cancer patients, environmental pollution and a drop in the local fishermen's catch." [*][38]

In addition to environmental and health costs at home, China's dependence on coal is cause for concern on a global scale. Due in large part to the emissions caused by burning coal, China is now the number one producer of carbon dioxide, responsible for a full quarter of the world's CO2 output.[*][39] According to a recent study, "even if American emissions were to suddenly disappear tomorrow, world emissions would be back at the same level within four years as a result of China's growth alone." [*][40] The country has taken steps towards battling climate change by pledging to cut its carbon intensity (the amount of CO2 produced per dollar of economic output) by about 40 per cent by 2020, compared to 2005 levels.[*][39] Reuters reports that "emissions and coal consumption will continue to rise through the 2020s, even though at a slower rate, barring a major intervention including a shift to cleaner burning gas from coal" - in other words, "meeting the carbon intensity target will require a significant change in trajectory for carbon emissions and coal consumption." [*][41] To that end, China has announced a plan to invest 2.3 trillion yuan ($376 billion) through 2015 in energy saving and carbon emission-reduction projects.[*][41]

China's first coal-fired power station employing the integrated gasification combined cycle (IGCC), which is a coal gasification process that turns coal into a gas before burning it, is planned to begin operations in 2009 at Tianjin near Beijing. Developed under a project called GreenGen, this $5.7 bn 650 MW plant will be a joint venture between a group of state-owned enterprises and Peabody Energy.[*][42] In addition to these coal gasification projects, it is worth noting that on average, China's coal plants work more efficiently than those in the United States, due to their relative youth.[*][7]

In September 2011, the Chinese government's Ministry of Environmental Protection announced a new emission standard for thermal power plants, for NOx and mercury, and a tightening of SO2 and soot standards. New coal power plants have a set date of the beginning of 2012 and for old power plants by mid-2014. They must also abide by a new limit on mercury by beginning of 2015. It is estimated such measures could bring about a 70% reduction in NOx emissions from power plants.[*][43]

In 2012, industrial conglomerate China Wanxiang Holdings signed a $1.25 billion deal with American company GreatPoint Energy to build a large-scale plant using Great-Point's catalytic hydromethanation process of coal gasification. The technology converts coal into natural gas and enables the recovery of contaminants in coal, petroleum coke and biomass as useful byproducts. Most importantly, nearly all of the CO2 produced in the process is captured as a pure stream suitable for sequestration or enhanced oil recovery.[*][44] The total project will cost an estimated $20 – 25 billion and will supply a trillion cubic feet of natural gas.[*][45] This represents a massive leap in the scale of domestic production for China, which last year produced only 107 billion cubic feet of natural gas.[*][46] The deal includes an equity investment of $420 million, the largest ever by a Chinese corporation into a venture-capital-funded U.S. company, according to industry tracker VentureSource.[*][44]

China is the first country with a single party government structure to take steps towards developing a nationwide Emissions Trading System.[*][47]

Beijing

China decided to close the last four coal-fired power and heating plants out of Beijing's municipal area, replacing them with gas-fired stations, in an effort to improve air quality in the capital. The four plants, owned by Huaneng Power International, Datang International Power Generation Co Ltd, China Shenhua Energy and Beijing Jingneng Thermal Power Co Ltd, had a total power generating capacity of about 2.7 gigawatts (GW). The first power plant closed in 2014, two other ones in 2015 and the last one will close in 2017.[48]

Coal mine fires

It is estimated that coal mine fires in China burn about 200 million kg of coal each year. Small illegal fires are frequent in the northern region of Shanxi. Local miners may use abandoned mines for shelter and intentionally set such fires. One study estimates that this translates into 360 million metric tons of carbon dioxide emissions per year, which is not included in the previous emissions figures.[49]

North China's Inner Mongolia Autonomous Region has announced plans to extinguish fires in the region by 2012. Most of these fires were caused by bad mining practices combined with bad weather. 200 million yuan (29.3 million USD) has been budgeted to this effect.[50]

3.3.4 Accidents and deaths

Main article: List of coal mining accidents in China

In 2003, the death rate per million tons of coal mined in China was 130 times higher than in the United States, 250 times higher than in Australia (open cast mines) and 10 times higher than the Russian Federation (underground mines). However the safety figures in the major state owned coal enterprises were significantly better. Even so, in 2007 China produced one third of the world's coal but had four fifths of coal fatalities.[51] It is also important to mention that China's coal mining industry resorts to forced labor according to a 2014 U.S. Department of Labor report on child labor and forced labor around the world,[52] and that these workers are all the more exposed to the dangers of such activities.

Pulmonary disease

While not directly attributable, many more deaths are resultant from dangerous emissions from coal plants. Chronic obstructive pulmonary disease (COPD), linked to exposure to fine particulates, SO_2, and cigarette smoke among other factors, accounted for 26% of all deaths in China in 1988.[54] A report by the World Bank in cooperation with the Chinese government found that about 750,000 people die prematurely in China each year from air pollution. Later, the government asked the researchers to soften the conclusions.[55]

Many direct deaths happen in coal mining and processing. In 2007, 1,084 out of the 3,770 workers who died were from gas blasts. Small mines (comprising 90% of all mines)

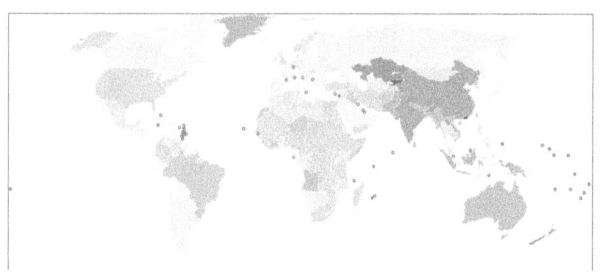

Disability-adjusted life year for chronic obstructive pulmonary disease per 100,000 inhabitants in 2004.[53]
no data
less than 110
110–220
220–330
330–440
440–550
550–660
660–770
770–880
880–990
990–1100
1100–1350
more than 1350

are known to have far higher death rates, and the government of China has banned new coal mines with a high gas danger and a capacity below 300,000 tons in an effort to reduce deaths a further 20% by 2010. The government has also vowed to close 4,000 small mines to improve industry safety.[56] A total of 2,657,230 people worked in state owned coal mines at the end of 2006.[57]

Accidents

The government has been cracking down on unregulated mining operations, which account for almost 80 percent of the country's 16,000 mines. The closure of about 1,000 dangerous small mines last year helped to cut in half the average number of miners killed, to about six a day, in the first six months of this year, the government has said. Major gas explosions in coal mines remain a problem, though the number of accidents and deaths have gradually declined year by year, the chief of the State Administration of Work Safety, Luo Lin, told a national conference in September.[58]

In the first nine months of 2009, China's coal mines had 11 major accidents with 303 deaths, with gas explosions the leading cause, according to the central government. Most accidents are blamed on failures to follow safety rules, including a lack of required ventilation or fire control equipment.[58]

Unofficial estimates often estimate death tolls at twice

the official number reported by the government.[59] Since 1949 over 250,000 coal mining deaths have been recorded.[60] However, since 2002, the death toll is gradually declining while the coal production is rapidly rising, doubling over this same period.

By year

A Chinese coal miner at the Jin Hua Gong Mine

Source: State Administration of Work Safety[63]

3.3.5   International opinions

See also: Debate over China's economic responsibilities for climate change mitigation

In October 2008, Greenpeace, World Wildlife Fund, and The Energy Foundation published The True Cost of Coal, a report that said that by-products of coal burning such as water pollution, air pollution and human costs such as mining deaths are costing China an additional 1.7 trillion yuan per year, or more than 7% of GDP. They recommended that China increase the price of coal by a tax of 23% to reflect the true costs of China's reliance on coal.[64]

Other commentators have pointed out that China has been taking a role as a leader in making use of coal as an electricity source *more* clean and responsible. For instance, the country built new ultra-supercritical coal plants (~44% efficiency) before the United States.[65] While the average efficiency of the coal fleet in China remains less than that of the US, the gap is quickly closing. China has required companies building new plants to retire an old plant for every new one built.[66]

3.3.6   See also

- China National Coal Group Corporation

- Shenhua Group

- Asian brown cloud

- *Other coal companies of China*

- Clean coal technology

Other countries

- Coal in the United States

3.3.7   References

[1] "Rise in China's Coal-fired Capacity in 2014, 2015 May Not Boost Thermal Coal Prices: UBS". Platts. May 5, 2014.

[2] http://www.wsj.com/articles/ chinas-coal-consumption-and-output-fell-last-year-1424956878

[3] "China Nov coal output down 2.7 pct at 320 mln T - stats bureau". Reuters. December 12, 2015.

[4] http://www.wsj.com/articles/ chinas-coal-consumption-and-output-fell-last-year-1424956878 Production 3959 TWh

[5] Feng, Hao (April 7, 2016). "China Puts an Emergency Stop on Coal Power Construction". The Diplomat.

[6] Myllyvirta, Lauri. "Comment: Why China's new coal mine moratorium matters". *Energy Desk*. Greenpeace.

[7] Cohen, Armond (April 21, 2014). "Learning from China: A Blueprint for the Future of Coal in Asia?". *The National Bureau of Asian Research*. Retrieved August 8, 2014.

[8] "Nuclear Power in China". *Country Briefings*. World Nuclear Association. 31 July 2010. Retrieved 2010-08-04.

[9] http://www.bp.com/content/dam/bp/pdf/ energy-economics/statistical-review-2015/ bp-statistical-review-of-world-energy-2015-coal-section. pdf

[10] Peter Fairley, *Technology Review*. Part I: China's Coal Future, January 5, 2007.

[11] "Country analysis briefs: China". Energy Information Administration. August 2006. Retrieved 2008-07-02.

[12] Jane Perlez and Yufan Huang (16 December 2015). "Mass Layoffs in China's Coal Country Threaten Unrest". *The New York Times*. Retrieved 17 December 2015. The coal industry is hurting nationwide, as coal prices have fallen nearly 60 percent since 2011, said Deng Shun, an analyst at ICIS C1 Energy, a consultancy based in Shanghai.

[13] IEA Key World Energy Statistics 2012, 2011, 2010, 2009, 2006 IEA coal production p. 15, electricity p. 25 and 27

[14] *The Age*. China coal shortage to continue. January 16, 2008.

[15] Bloomberg. China Shuts More Coal Power Plants; Warns on Shortage (Update1). June 8, 2008.

[16] "China's 7 Coal Mining Companies Realized Production Capacity of 100 Mln Tonnes in 2011". China Mining Association. 2012-02-01. Retrieved 2012-06-03.

[17] Le, Reggie (2012-04-05). "China's Jizhong Energy mines 31 million mt of coal, up 10% on year". *Platts*. Retrieved 2012-06-03.

[18] http://www.nytimes.com/2015/11/04/world/asia/china-burns-much-more-coal-than-reported-complicating-climate-talks.html

[19] "China's largest open-pit coal mine ready for production". Xinhua News Agency. October 19, 2008. Retrieved 2010-08-04.

[20] "China's Growth Leads to Problems Down the Road" "Mongolian coal production has exploded —up 37 percent to 637 million tons last year alone, with an additional 15 percent increase expected this year." article by Michael Wines in *The New York Times* August 27, 2010, accessed August 28, 2010

[21] Jin, Tony (October 27, 2009). "China Consumes 9% More Coal through September". *The China Perspective*. Retrieved 2010-08-04.

[22] "Coal and Peat in China, People's Republic of in 2007". International Energy Agency (IEA). Retrieved 2010-08-04.

[23] Other Transformation refers to an energy transformation process not in the preceding list of electricity, industry, or heat. For the case of coal, this is likely to include losses, own use, gains, or liquefaction. Reference: .

[24] http://theenergycollective.com/michael-davidson/335271/china-s-electricity-sector-glance-2013

[25] http://marketrealist.com/2014/10/coal-is-losing-its-market-share-in-chinas-electricity-generation/

[26] *Technology Review*. Part II: China's Coal Future, "To prevent massive pollution and slow its growing contribution to global warming, China will need to make advanced coal technology work on an unprecedented scale."

[27] Wikinvest:China'{}s Coal Power Pollution.

[28] Ma, Damien. "China's Coming Decade of Natural Gas". *Asia's Uncertain LNG Future*. November 2013. Retrieved 8 August 2014 from http://www.nbr.org/publications/element.aspx?id=711.

[29] Robert B. Finkelman, Harvey E. Belkin, and Baoshan Zheng. Health impacts of domestic coal use in China. *Proc Natl Acad Sci U S A*. 1999 March 30; 96(7): 3427–3431.

[30] Environmental Health Perspectives. Household Air Pollution from Coal and Biomass Fuels in China: Measurements, Health Impacts, and Interventions. Received July 3, 2006; Accepted February 27, 2007.

[31] FT.com / Asia-Pacific / China - Australia loses market share in China's coal

[32] "China's coal imports fall nearly half in 12 months as anti-pollution drive bites" 'Guardian/Reuters *13 April 2015*

[33] http://www.reuters.com/article/2014/09/21/us-un-climatechange-carbon-idUSKBN0HG0QA20140921

[34] *The Christian Science Monitor*. New coal plants bury 'Kyoto'. December 23, 2004.

[35] "Country analysis briefs: China". Energy Information Administration. August 2006. Retrieved 2008-07-11.

[36] Spencer,Richard "Pollution kills 750,000 in China every year" *The Telegraph UK*, 4 July 2007

[37] Andrews-Speed, Philip (November 2014). "China's Energy Policymaking Processes and Their Consequences". *The National Bureau of Asian Research Energy Security Report*. Retrieved December 24, 2014.

[38] "South China town unrest cools after dialogue" *Associated Foreign Press*, 23 December 2011

[39] Bawden, Tom "China agrees to impose carbon targets by 2016" *The Independent*, 21 May 2013

[40] Muller, Elizabeth "China Must Exploit Its Shale Gas" *The Telegraph*, 12 April 2013

[41] Wynn, Gerard "China's carbon goal shows coal growth has peaked" *Reuters*, 7 August 2013

[42] "China's first carbon capture & storage plant to be operational by 2009". Power Engineering International. 2007-12-31. Retrieved 2008-01-13.

[43] "Chinese government demand coal companies begin to pay for bad air". Greenpeace East Asia. 2011-09-26. Retrieved 2011-09-26.

[44] Kolodny, Lora "Bluer Skies for Shanghai?", *Wall Street Journal Venture Capital Dispatch*, 20 February 2012

[45] Daniels, Steve "A Chicago company brings power to the People's Republic", *Crain's Chicago Business*, 12 September 2012

[46] "China's natural gas consumption up 13% in 2012" *China Knowledge Newswires*, 29 January 2013

[47] http://www.ieta.org/assets/Reports/EmissionsTradingAroundTheWorld/edf_ieta_china_case_study_september_2013.pdf

[48] Chen, Kathy; Tom Miles (22 May 2015). "Beijing promises coal-free power by 2017 to fight pollution". Reuters. Retrieved 2015-11-25.

[49] Mines and Communities Website. A Burning Issue. February 14, 2003.

[50] Xinhua. N China to put out some coalfield fires by 2012. 2010-06-04

[51] World Investment Report 2007: Transition Corporations, Extractive Industries United Nations Conference on Trade and Development page 149

[52] List of Goods Produced by Child Labor or Forced Labor

[53] "WHO Disease and injury country estimates". *World Health Organization*. 2009. Retrieved Nov 11, 2009.

[54] China and Coal. Archived November 16, 2007, at the Wayback Machine.

[55] *Financial Times*. 750,000 a year killed by Chinese pollution. July 2, 2007.
released version of the report:

[56] Xinhua. China to ban small coal mines for improving pit safety record. August 15, 2008.

[57] International Energy Agency. Cleaner Coal in China. Copyright 2009.

[58] "42 Reported Dead, and 66 Trapped, in China Mine Accident" by the Associated Press, via *The New York Times*. November 21, 2009. Retrieved 2009-11-21.

[59] World Socialist Website. China's coal mining deaths spiral. August 3, 2002.

[60] *International Herald Tribune*. Chinese coal industry in need of a helping hand

[61] http://news.xinhuanet.com/english2010/china/2011-02/25/c_13750546.htm

[62] http://chinadigitaltimes.net/2012/09/20-die-in-coal-mine-plunge/

[63] Mines and Communities Website. China and US coal disasters. 7th January 2006.

[64] Greenpeace web site. China's Coal Crisis. October 27, 2008.

[65] Technology Frontier Article

[66] *The New York Times*. China Outpaces U.S. in Cleaner Coal-Fired Plants. May 10, 2009.

3.3.8   External links

Organizations

- Worldwide Coal Production In China (Live-Counter)
- China Coal Society
- Coalbed Methane Committee - China Coal Society
- China Coal Industry Development Research Center
- China Coal Industry Network - Chinese coal industry's information, policy, science and technology, statistics and other information.
- China Coal Industry Association
- Statecoal.com - National Coal Network
- General Administration of Coal Geology
- State Coal Mine Safety Supervision
- National Coal Mine Safety Supervision Bureau
- Coal Mining Production Safety Information Network
- China Coal Research Institute - Coalfield Geology and exploration, coal mining, coal mine safety, coal mining machinery, coal washing, coal mine environmental protection, pipeline transporting coal and other professional applied & basic research.
- Beijing Research Institute of Coal Chemistry
- List of coal research institutes in China

Articles

- Peter Fairley, *Technology Review*. Part I: China's Coal Future, January 5, 2007.
- Peter Fairley, *Technology Review*. Part II: China's Coal Future, January 5, 2007.
- China to enhance coal industry restructuring
- The True Cost of Coal: Greenpeace China Report on China's Coal Crisis

3.4   Tianjin Climate Exchange

Tianjin Climate Exchange (TCX) is a domestic carbon market cap-and-trade scheme exchange. Jeff Huang is assistant chairman of Tianjin Climate Exchange and vice-president of Chicago Climate Exchange.

It is China's first integrated exchange for trading of environmental financial instruments

TCX is a joint venture between Chicago Climate Exchange, the municipal government of Tianjin and the asset management unit of PetroChina, the country's largest oil and gas producer.

Cap-and-trade schemes are programs under which member companies commit to lowering their greenhouse gas emissions by a certain amount in a certain period of time and trade carbon credits generated by this. As China does not have a national cap on emissions, any such scheme would be voluntary, similar to the situation in the US when the Chicago Climate Exchange launched in 2003.

3.4.1 History

On September 25 2008, Tianjin Climate Exchange, co-established by CNPC Assets Management Co., Ltd. (holding a 53% stake), Tianjin Property Rights Exchange (北方产权交易市场) (holding a 22% stake), and Chicago Climate Exchange (CCX) (holding a 25% stake), was unveiled in the Tianjin Binhai New Area.

At the request of the State Council, Tianjin Climate Exchange is established as China's first comprehensive platform for trading carbon credits under the Clean Development Mechanism, and will promote environmental protection and emission reduction by means of market and financial measures.

Tianjin Climate Exchange has the following goals: to help enterprises cost-effectively reduce emissions of pollutants, such as sulfur dioxide, chemical oxygen demand, etc.; to help enterprises achieve maximum energy efficiency at minimum cost; to help enterprises manage environmental risks and meet increasing disclosure requirements; and to provide enterprises with integrated international emissions market access and experience.

In 2006, Tianjin Binhai New Area was designed by the State Council of the PRC as the national experimental zone for comprehensive reforms related to financial innovation, land and administrative management.

China's Eleventh Five-Year Plan (2006-10) called for cutting energy consumption per unit of GDP up to 20 percent by 2010 while reducing major pollutants, such as sulfur dioxide (SO2) by 10 percent.[3]

3.4.2 Tianjin Property Rights Exchange

TPRE was launched in 1994, under government approval. It is a government agent under the charge of Tianjin SASAC and is the only appointed exchange authorized by Tianjin SASAC for state-owned assets and equities transaction. It is one of three national institutions permitted by SASAC to transact assets and equities of SOEs under control of central government.[2]

3.4.3 See also

- Clean Development Mechanism

- Beijing Environmental Exchange

- Richard Sandor

3.4.4 References

- http://www.chicagoclimatex.com/news/press/release_20081223_CLE_announcement.pdf

- http://www.cnpc.com.cn/eng/press/newsreleases/CSR/d1495066-3c79-482a-a36b-88b834cf3f78.htm

- http://www.chinadaily.com.cn/bizchina/2008-10/27/content_7144537.htm

3.4.5 External links

- official website

Chapter 4

Pollution in China

4.1 Center for Legal Assistance to Pollution Victims

The **Center for Legal Assistance to Pollution Victims or CLAPV** (simplified Chinese: 污染受害者法律帮助中心) at the China University of Political Science and Law is a legal-aid office, training center, and one of the most effective environmental groups in China.[*][1]

CLAPV was established in October 1998 by Professor Wang Canfa. It was approved by China University of Political Science & Law and registered with the Chinese Judicial Ministry. According to Professor Wangcafa, "China has many laws and regulations regarding the environment, but the situation just gets worse, because they are often not implemented."

Goals of CLAPV include promoting "enforcement of environmental law," and to "tell the public how to respond when your rights are violated." [*][2]

CLAPV had answered more than 10,000 calls for assistance to its hotline and pursued over 100 cases, some with as many as 1,700 plaintiffs.[*][3]

4.1.1 References

Notes

[1] "NRDC Backgrounder: China Environmental Law Group Partners with Leading U.S. Environmental Organization". Nrdc.org. 2006-02-14. Retrieved 2012-01-25.

[2] Pollution Victims Start to Fight Back 2000.050.6 (Retrieved on November 25, 2011)

[3] Wang Canfa | Heroes of the Environment. *Time*. 2007.10. (Retrieved on November 25, 2011)

4.1.2 External links

- CLAPV official website, accessed 27 February 2013

- CLAPV official website, English page, accessed 27 February 2013

4.2 China Pollution Map Database

Since its establishment in May 2006, the Institute of Public & Environmental Affairs (IPE), a registered non-profit organization based in Beijing, China, has developed the China Pollution Map Database to monitor corporate environmental performance, pinpoint geographical locations of pollution sources and to act as an informational platform on regional pollution status, such as water and air quality, and pollutant discharge rankings. This publicly available information resource brings together over 97,000 environmental supervision records from government departments, at all levels and regions, throughout mainland China. These records, dating back as far as 2004, allow for the expansion of environmental information disclosure, enabling communities to fully understand the hazards and risks in the surrounding environment, thus promoting widespread public participation in environmental governance.

With the improvements made to the IPE website and thus 'China Pollution Map Database,' over the last two years the number of official government-sourced violation records, from all regions in China, added to the 'China Pollution Map Database,' has grown by over 40,000; each having its own particular circumstances and need for a prompt and effective resolution.

This important publicly available resource allows stakeholders, at all levels, to be informed, providing opportunities for individual and community inclusion, as well as NGO and media engagement. The IPE hopes this societal supervision of corporate and regional environmental performance and the external pressure on government will promote greater efforts towards enforcement and compliance, and in turn will provide a safer, cleaner and healthier environment for all.

When developing and updating the 'China Pollution Map

Database,' the IPE pays special attention to the annually published, government authored, *List of Key State Monitored Enterprises**[1] which according to set selection principles and methods, identifies those companies that occupy 65% of China's industrial discharge volume, in traditionally heavy polluting industries. Through, often difficult, research and field visits, the IPE aims to continue the geographical positioning of the location of these companies.

In continuation of this program, the IPE intends to continue arduously collecting, organizing and disseminating this environmental data, and information on corporate violations from around China.

4.2.1 See also

- Pollution in China

4.2.2 References

[1] *List of Key State Monitored Enterprises*

4.2.3 External links

- Institute of Public & Environmental Affairs

- China Pollution Map Database

- *A Road Map to Blue Skies – China's Atmospheric Pollution Source Positioning Report*

4.3 JSYU UAV

JSYU UAVs are Chinese UAVs developed by Jiang-Su You-Tu (meaning top-notch imagery) Spatial Information Science and Technology Co., Ltd (JSYU, 江苏优图空间信息科技公司) for pollution surveillance and agricultural missions.

4.3.1 UV-I

UV-I is the first UAV developed by JSYU for environmental surveillance missions, mainly for aerial pollution surveillance, particularly for pollutants with size of PM 2.5. Another mission is for agricultural mission, mainly for surveillance of pest and disease affected crops. To achieve such mission, the UV-I is equipped with hyperspectral imaging system. However, the expensive price tag for such system has prevented its wide use, because foreign hyperspectral imaging systems cost between 10 to 20 million ¥, and for the

UAV platform, the price is at least three million ¥. The reason for such high price is because the weight of the hyperspectral imaging system is heavy and thus requiring larger UAV with higher price tag to carry it. As a result, JSYU decided to develop its own hyperspectral imaging system and UAV platform to carry it, which would be much cheaper then imported foreign counterparts. The result of JSYU's effort of indigenous development is UV-I, which carries a light weight hyperspectral imaging system indigenously developed by JSYU.*[1]

Development of UV-I begun in September 2013 and maiden flight was completed at the beginning of 2014. UV-I is a fixed-wing UAV in conventional layout with high wing configuration and tricycle landing gear. Propulsion is provided by a two-blade propeller driven tractor engine mounted in the nose. The wing can be easily removed/installed for storage and rapid development. One of the greatest challenge is that the UAV can only carry a maximum of 25 kg, and most hyperspectral imaging system is heavier than this limit and thus cannot be carried. Dr. Zhang Li-Fu (张立福) of Institute of Remote Sensing and Digital Earth (中国科学院遥感与数字地球研究所) of Chinese Academy of Sciences CAS developed the light weight compact system needed, with weight reduced to 10 kg, and further development will reduce the weight by another half to 5 kg.*[1] The hyperspectral imaging system is manufactured at JSYU's Beijing site, while the UAV platform is manufactured at JSYU's Yangzhou site. The price of the UAV platform is only a third of those imported ones, at one million ¥, and the overall system is around seventy or eighty percent cheaper than the imported ones.*[1] Specification:*[1]

- Wingspan (m): 3

- Payload (kg): 25

- Endurance (hr): 8

- Normal operating altitude (km): 1

4.3.2 UV-II

UV-II is the development of UV-I based on experience learned from UV-I, and it is fielded in mid 2014. UV-II is constructed of honeycomb carbon fiber and glass fiber reinforced plastic, thus increased strength of the airframe while ducing the weight by a third.*[2] The mission capability has expanded from the original agricultural and environmental surveillance to aerial photography, mineral surveying and other missions, and UV-II can operate in light rain. UV-II is almost identical to its predecessor UV-I in appearance, and the only obvious external visual difference between the two is that the tailplanes of UV-II have winglets. The landing gear of UV-II is constructed for CZ series aluminum alloy to provide strength to withstand 50 kg force while reducing weight. Like its predecessor UV-I, UV-II can be easily assembled in the field for rapid deployment.[*][2] Specification:[*][2]

- Wingspan (m): 2.96

- Length (m): 2.4

- Wing area (sq m): 1.2

- Height (m): 0.5

- Max speed (km/hr): 170

- Cruise speed (km/hr): 120

- Ceiling (km): 5.5

- Rate of climb (m/s): 10

- Typical operation radius (km): 140

- Endurance (hr): 3

- Range (km): 200 – 400

- Remote control radius (km): 30 – 50

- Navigation: GPS

- Empty weight (kg): 10

- Payload (kg): 15

- Max take-off weight (kg): 30

- Max fuel capacity (kg): 10

- Max g-force allowed: 2

- Temperature allowed for operation (°C): −30 to +60

- Wind speed allowed for cruise flight (m/s): 13

- Max wind scaled allowed for launch and recovery: 5

- Launch: taxiing or catapult

- Recovery: taxiing or parachute

- Take-off & landing distance (m): 50

- Deployment time (hr): 0.5 (from storage to airborne)

- Minimum time between two consecutive sorties (min): 10

4.3.3   See also

List of unmanned aerial vehicles of the People's Republic of China

4.3.4   References

[1]   "UV-I". Retrieved Jun 12, 2014.

[2]   "UV-II". Retrieved Jun 12, 2014.

4.4   Pollution in China

Beijing air on a 2005-day after rain (left) and a smoggy day (right)

Pollution is one aspect of the broader topic of environmental issues in China. Various forms of pollution have increased as China has industrialized, which has caused widespread environmental and health problems.[*][1][*][2]

4.4.1   Pollution is a problem

Soil contamination

Further information: Soil contamination § People's Republic of China

The immense growth of the People's Republic of China since the 1980s has resulted in increased soil pollution. The State Environmental Protection Administration believes it to be a threat to the environment, food safety and sustainable agriculture. 38,610 square miles (100,000 km^2) of China's cultivated land have been polluted, with contaminated water being used to irrigate a further 31.5 million miles (21,670 km^2.) and another 2 million miles (1,300 km^2) have been covered or destroyed by solid waste. In total, the area accounts for one-tenth of China's cultivatable land, and is not known as the first time mostly in economically developed areas. An estimated 19 million tonnes of grain are contaminated by heavy metals every year, causing direct losses of 29 billion yuan (US$2.57 billion).

Waste

As China's waste production increases, insufficient efforts to develop capable recycling systems have been attributed to a lack of environmental awareness.[*][3] In 2012 the waste generation in China was 300 million tons (229.4 kg/cap/yr).[*][4]

A ban came into effect on June 1, 2008 that prohibited all supermarkets, department stores and shops throughout China from giving out free plastic bags.[*][5] Stores must clearly mark the price of plastic shopping bags and are banned from adding that price onto the price of products. The production, sale and use of ultra-thin plastic bags - those less than 0.025 millimeters (0.00098 in) thick - are also banned. The State Council called for "a return to cloth bags and shopping baskets." [*][6] This ban, however, does not affect the widespread use of paper shopping bags at clothing stores or the use of plastic bags at restaurants for takeout food. A survey by the International Food Packaging Association found that in the year after the ban was implemented, 10% fewer plastic bags found their way into the garbage.[*][7]

Electronic waste   Main article: Electronic waste in China

In 2011, China produced 2.3 million tons of electronic waste.[*][8] The annual amount is expected to increase as the Chinese economy grows. In addition to domestic waste production, large amounts of electronic waste are imported from overseas. Legislation banning importation of electronic waste and requiring proper disposal of domestic waste has recently been introduced, but has been criticized as insufficient and susceptible to fraud. There have been local successes, such as in the city of Tianjin where 38,000 tons of electronic waste were disposed of properly in 2010, but much electronic waste is still improperly handled.[*][9]

Industrial pollution

Air pollution caused by industrial plants

In 1997, the World Bank issued a report targeting China's policy towards industrial pollution. The report stated that

"hundreds of thousands of premature deaths and incidents of serious respiratory illness [have been] caused by exposure to industrial air pollution. Seriously contaminated by industrial discharges, many of China's waterways are largely unfit for direct human use". However, the report did acknowledge that environmental regulations and industrial reforms had had some effect. It was determined that continued environmental reforms were likely to have a large effect on reducing industrial pollution.[*][10]

In a 2007 article about China's pollution problem, the New York Times stated that "Environmental degradation is now so severe, with such stark domestic and international repercussions, that pollution poses not only a major long-term burden on the Chinese public but also an acute political challenge to the ruling Communist Party." The article's main points included:[*][11]

1. According to the Chinese Ministry of Health, industrial pollution has made cancer China's leading cause of death.

2. Every year, ambient air pollution alone killed hundreds of thousands of citizens.

3. 500 million people in China are without safe and clean drinking water.

4. Only 1% of the country's 560 million city dwellers breathe air considered safe by the European Union, because all of its major cities are constantly covered in a "toxic gray shroud". Before and during the 2008 Summer Olympics, Beijing was "frantically searching for a magic formula, a meteorological deus ex machina, to clear its skies for the 2008 Olympics."

5. Lead poisoning or other types of local pollution continue to kill many Chinese children.

6. A large section of the ocean is without marine life because of massive algal blooms caused by the high nutrients in the water.

7. The pollution has spread internationally: sulfur dioxide and nitrogen oxides fall as acid rain on Seoul, South Korea, and Tokyo; and according to the Journal of Geophysical Research, the pollution even reaches Los Angeles in the USA.

8. The Chinese Academy of Environmental Planning in 2003 produced an unpublished internal report which estimated that 300,000 people die each year from ambient air pollution, mostly of heart disease and lung cancer.

9. Chinese environmental experts in 2005 issued another report, estimating that annual premature deaths attributable to outdoor air pollution were likely to reach 380,000 in 2010 and 550,000 in 2020.

10. A 2007 World Bank report conducted with China's national environmental agency found that "...outdoor air pollution was already causing 350,000 to 400,000 premature deaths a year. Indoor pollution contributed to the deaths of an additional 300,000 people, while 60,000 died from diarrhea, bladder and stomach cancer and other diseases that can be caused by water-borne pollution." World Bank officials said "China' s environmental agency insisted that the health statistics be removed from the published version of the report, citing the possible impact on 'social stability'".

A draft of a 2007 combined World Bank and SEPA report stated that up to 760,000 people died prematurely each year in China because of air and water pollution. High levels of air pollution in China's cities caused to 350,000-400,000 premature deaths. Another 300,000 died because of indoor air of poor quality. There were 60,000 premature deaths each year because of water of poor quality. Chinese officials asked that some of results should not be published in order to avoid social unrest.[12]

China has made some improvements in environmental protection during recent years. According to the World Bank, 'China is one of a few countries in the world that have been rapidly increasing their forest cover. It is managing to reduce air and water pollution.[13]

Vennemo et al. in a 2009 literature review in *Review of Environmental Economics and Policy* noted the wide discrepancy between the reassuring view in some Chinese official publications and the exclusively negative view in some Western sources. The review stated that "although China is starting from a point of grave pollution, it is setting priorities and making progress that resemble what occurred in industrialized countries during their earlier stages of development." Environmental trends were described as uneven. Quality of surface water in the south of China was improving and particle emissions were stable. But NO_2 emissions were increasing rapidly and SO_2 emissions had been increasing before decreasing in 2007, the last year for which data was available.[14]

Conventional approaches to air quality monitoring are based on networks of static and sparse measurement stations. However, there are drivers behind current rises in the use of low-cost sensors for air pollution management in cities.[15]

Water pollution

Main article: Water pollution in China

The water resources of China are affected by both severe water shortages and severe water pollution. An increas-

ing population and rapid economic growth as well as lax environmental oversight have increased water demand and pollution. In response, China has taken measures such as rapidly building out the water infrastructure and increased regulation as well as exploring a number of further technological solutions.

Air pollution

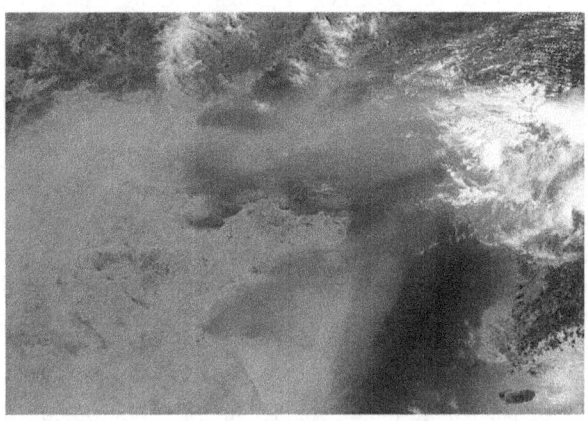

Thick haze blown off the Eastern coast of China, over Bo Hai Bay and Yellow Sea. The haze might result from urban and industrial pollution.

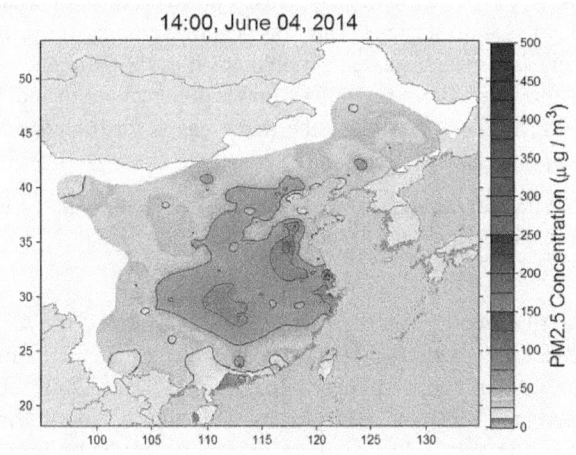

Map of PM2.5 pollution over China from April to August 2014

Air pollution has become a major issue in China, and poses a threat to Chinese public health. Coal combustion generates particulate matter also known as "PM". Currently, Beijing is suffering from PM2.5, which is a particulate matter with diameter of 2.5 micrometers or less. According to the U.S. Environmental Protection Agency, such fine particles can cause asthma, bronchitis, and acute and chronic respiratory symptoms such as shortness of breath and painful breathing, and may also lead to premature death.[16] The Telegraph reported a case of an 8-year-old girl who had contracted lung cancer, becoming the youngest victim of lung cancer in China.[17] Doctors pointed out that the likely cause was exposure of air pollution, specifically fine particulates from vehicles. The case has gathered large national public attention and also international attention.

Zhong Nanshan, the president of the China Medical Association, warned in 2012 that air pollution could become China's biggest health threat. Lung cancer and cardiovascular disease were increasing because of factory and vehicle air pollution and tobacco smoking. Lung cancer was two to three times more common in cities than in the countryside despite similar rates of tobacco smoking. Zhong stated that while transparency had increased in recent years much more information was needed, and called for detailed epidemiological research. He questioned official data stating that air pollution was decreasing. Until recently the governmental air quality index did not include ozone and PM2.5, despite these being the most dangerous to human health.[18] Measurements in January 2013 showed that levels of air pollution, as measured by the density of particulate matter smaller than 2.5 micrometres in size, were beyond index – higher than the maximum 755 μg the US Embassy's equipment can measure.[19] Smog from mainland China has been observed to reach as far as California.[20]

Sulfur dioxide emissions increased until 2006, after which they began to decline. This was accompanied by improvements on several related variables such as the frequency of acid rainfall. The adoption by power plants of sulfur reducing technology was likely the main reason for the reduced SO_2 emissions.[21]

Large scale use of formaldehyde in construction and furniture also contributes to indoor air pollution.[22]

Particulates According to the World Bank, the Chinese cities with the highest levels of particulate matter in 2004 of those studied were Tianjin, Chongqing, and Shenyang.[23] In 2012 stricter air pollution monitoring of ozone and PM2.5 were ordered to be gradually implemented so that by 2015 all but the smallest cities would be included. State media acknowledged the role of environmental campaigners in causing this change. On one microblog service more than a million mostly positive comments were posted in less than 24 hours although some wondered if the standards would be effectively enforced.[24]

The US embassy in Beijing regularly posts automated air quality measurements at @beijingair on Twitter. On 18 November 2010, the feed described the PM2.5 measurement as "crazy bad" after registering a reading in excess of 500 for the first time. This description was later changed

to "beyond index",[25] a level which recurred in February, October, and December 2011.[26][27][28]

In June 2012, following strongly divergent disclosures of particulate levels between the Observatory and the US Embassy, Chinese authorities asked foreign consulates to stop publishing "inaccurate and unlawful" data.[29] Controversy arose when U.S. Embassy declared Beijing air as "very unhealthy" on 5 June; underlying data showed 199 micrograms of particulate matter. In contrast, readings from the Beijing Municipal Environmental Protection Bureau declared Beijing air as "good"; its data showed levels between 51 and 79 micrograms for the corresponding period.[30] Officials said it was "not scientific to evaluate the air quality of an area with results gathered from just only one point inside that area", and asserted that official daily average PM2.5 figures for Beijing and Shanghai were "almost the same with the results published by foreign embassies and consulates".[29]

By January 2013 the pollution had worsened with official Beijing data showing an average figure over 300 and readings of up to 700 at individual recording stations while the US Embassy recorded over 755 on January 1 and 800 by January 12.[31][32]

On October 21, 2013, record smog closed the Harbin Airport along with all schools in the area. Daily particulate levels of more than 50 times the World Health Organisation recommended daily level were reported in parts of the municipality.[33]

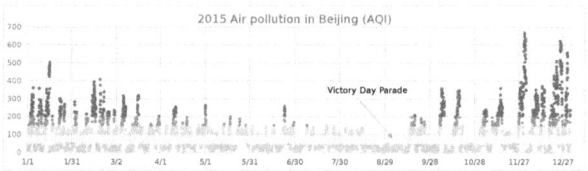

2015 Air pollution in Beijing as measured by Air Quality Index (AQI)
Severely Polluted
Heavily Polluted
Moderately Polluted
Lightly Polluted
Good
Excellent

Government's response to the air pollution In an attempt to reduce air pollution, the Chinese government has made the decision to enforce stricter regulations. After record-high air pollution in northern China in 2012 and 2013, the State Council issued an Action Plan for the Prevention and Control of Air Pollution in September 2013. This plan aims to reduce air pollution by over 10% from

2012 to 2017.*[34] The most prominent government response has been in Beijing.*[35] As the capital of China, it is suffering from high levels of air pollution. According to Reuters, in September 2013, the Chinese government published the plan to tackle air pollution problem on its official website.*[36] The main goal of the plan is to reduce coal consumption by closing polluting mills, factories and smelters and switching to other eco-friendly energy sources.*[35]

On 20 August 2015, ahead of the 70th anniversary celebrations of the end of World War II, the Beijing government shut down industrial facilities and reduced car emissions in order to achieve a "Parade Blue" sky for the occasion. This action resulted in PM2.5 concentration better than the 35 mg/m^3 national air quality standard, according to data from Beijing Municipal Environmental Protection Monitoring Centre (BMEMC). The restrictions resulted in an average Beijing PM2.5 concentration of 19.5 mg/m^3, the lowest that had ever been on record in the capital.*[37]

China's strategy has been largely focusing on the development of other energy sources such as nuclear, hydro and compressed natural gas. The latest plan entails closing the outdated capacity of the industrial sectors like iron, steel, aluminum and cement and increasing nuclear capacity and other non-fossil fuel energy. It also includes an intention to stop approving new thermal power plants and to cut coal consumption in industrial areas.

Four-color alert system Beijing launched four-color alert system in 2013. It is based on the air quality index (AQI), which indicates how clean or polluted the air is.*[38]

The Beijing government revised their four color alert system at the start of 2016, increasing the levels of pollution required to trigger orange and red alerts. The change was introduced to standardise the alert levels across four cities including Tianjin and four cities in Hebei, and perhaps in direct response to the red alerts issues the previous December.*[39]*[40]

4.4.2   Pollutants

Lead

Lead poisoning was described in a 2001 paper as one of the most common pediatric health problems in China. A 2006 review of existing data suggested that one-third of Chinese children suffer from elevated serum lead levels. Pollution from metal smelters and a fast-growing battery industry has been responsible for most cases of particularly high lead levels. In 2011, there were riots in the Zhejiang Haijiu Battery Factory from angry parents whose children received perma-

nent neurological damage from lead poisoning. The central government has acknowledged the problem and has taken measures such as suspending battery factory production, but some see the response as inadequate and some local authorities have tried to silence criticisms.*[41]

A literature review of academic studies on Chinese children's blood lead levels found that the lead levels declined when comparing the studies published during 1995-2003 and 2004-2007 periods. Lead levels also showed a declining trend after China banned lead in gasoline in 2000. Lead levels were still higher than those in developed nations. Industrial areas had higher levels than suburban areas, which had higher levels than urban areas. Controlling and preventing lead poisoning was described as a long term mission.*[42]

Persistent organic pollutants

China is a signatory nation of the Stockholm Convention, a treaty to control and phase out major persistent organic pollutants (POP). A plan of action for 2010 includes objectives such as eliminating production, import and use of the pesticides covered under the convention, as well as an accounting system for PCB containing equipment. For 2015, China plans to establish an inventory of POP contaminated sites and remediation plans.*[43] Since May 2009, this treaty also covers polybrominated diphenyl ethers and perfluorooctanesulfonic acid. Perfluorinated compounds are associated with altered thyroid function and decreased sperm count in humans.*[44] China faces a big challenge in controlling and eliminating POPs, since they often are cheaper than their alternatives, or are unintentionally produced and then released into the environment to save on treatment costs.

Dust

The Yellow Dust or Asian dust is a seasonal dust cloud which affects North East Asia during late winter and springtime. The dust originates in the deserts of Mongolia, northern China and Kazakhstan where high-speed surface winds and intense dust storms kick up dense clouds of fine, dry soil particles. These clouds are then carried eastward by prevailing winds and pass over Northern China into Korea and Japan.

Desertification has intensified in China. 1,740,000 square kilometres of land is classified as "dry", and desertification disrupts the lives of 400 million people and causes direct economic losses of 54 billion yuan ($7 billion) a year, SFA figures show.*[45] Sulfur (an acid rain component), soot, ash, carbon monoxide, and other toxic pollutants including heavy metals (such as mercury, cadmium, chromium, arsenic, lead, zinc, copper) and other carcinogens, often accompany the dust storms, as well as viruses, bacteria, fungi, pesticides, antibiotics, asbestos, herbicides, plastic ingredients, combustion products and hormone mimicking phthalates.*[46]

Other pollutants

In 2010 49 employees at Wintek were poisoned by n-hexane in the manufacturing of touchscreens for Apple products.*[47]

In 2013, it was revealed that portions of the country's rice supply were tainted with the toxic metal cadmium.*[48]

4.4.3 Impact of pollution

A 2006 Chinese green gross domestic product estimate stated that pollution in 2004 cost 3.05% of the nation's economy.*[49]

A 2007 World Bank and SEPA report estimated the cost of water and air pollution in 2003 to 2.68% or 5.78% of GDP depending on if using a Chinese or a Western method of calculation.*[50]

A 2009 review stated a range of 2-10% of GDP.*[14]

A 2012 study stated that pollution had little effect on economic growth which in China's case was largely dependent on physical capital expansion and increased energy consumption due to the dependency on manufacturing and heavy industries. China was predicted to continue to grow using energy-inefficient and polluting industries. While growth may continue, the rewards of this growth may be opposed by the harm from the pollution unless environmental protection is increased.*[51]

A 2015 report from the University of California at Berkeley estimated that 1.6 million people in China die each year from heart, lung and stroke problems because of polluted air.*[52]

4.4.4 Criticisms of government environmental policies

Critics point to the government's lack of willingness to protect the environment as a common problem with China's environmental policies. Even in the case of the latest plan, experts are skeptical about its actual influence because of the existence of loopholes. This is because economic growth is still the primary issue for the government, and overrides environmental protection.

However, if the measures to cut coal usage were applied

strictly, it would also mean dismantling of the local economy that is highly reliant on heavy industry. The Financial Times interviewed a worker who stated, "if this steel mill didn't exist, we wouldn't even have anywhere to go to eat. Everything revolves around this steel factory – our children work here." *[53]

4.4.5 Pollution ratings

As of 2004:

- The top five environmentally friendly cities: Haikou, Zhuhai, Zhanjiang, Guilin, Beihai*[54]

- The top five cities with most effective pollution controls: Nantong, Lianyungang, Shenyang, Suzhou, Fuzhou*[54]

- The 10 cities with worst air quality: Linfen, Yangquan, Datong, Shizuishan, Sanmenxia, Jinchang, Shijiazhuang, Xianyang, Zhuzhou, Luoyang*[54]

According to the National Environmental Analysis released by Tsinghua University and The Asian Development Bank in January 2013, 7 of the 10 most air polluted cities in the world are in China, including Taiyuan, Beijing, Urumqi, Lanzhou, Chongqing, Jinan and Shijiazhuang.*[55]

4.4.6 See also

- 2009 Chinese lead poisoning scandal

- 2013 Eastern China smog

- 2013 Northeastern China smog

- Automotive industry in China

- China Energy Conservation Investment Corporation

- China Pollution Map Database

- Climate change in China

- Environment of China

- Environmental issues in China

- List of power stations in China

- Low-carbon economy

- Peak oil

- Renewable energy in China

- List of countries by energy consumption and production

- Category:Energy by country

- Haze

- Smog

4.4.7 References

[1] "The Most Polluted Places On Earth". CBS News. 2010-01-08. Retrieved 2013-01-21.

[2] "Air Pollution Grows in Tandem with China's Economy". NPR. Retrieved 2013-01-21.

[3] Violet Law (July 28, 2011). "As China's prosperity grows, so do its trash piles". *The Christian Science Monitor*. Retrieved 29 July 2011.

[4] Waste Atlas (2012). Country Data: CHINA

[5] Prashant, Kumar,; Lidia, Morawska,; Claudio, Martani,; George, Biskos,; Marina, Neophytou,; Di, Sabatino, Silvana; Margaret, Bell,; Leslie, Norford,; Rex, Britter, (2015-02-01). "The rise of low-cost sensing for managing air pollution in cities". *eprints.qut.edu.au*. Retrieved 2016-01-19.

[6] "China bans free plastic shopping bags", AP Press via the *International Herald Tribune*, January 9, 2008

[7] David Biello, Scientific American, Does Banning Plastic Bags Work?, August 13, 2009.

[8] BONN. "Urgent Need to Prepare Developing Countries for Surge in E-wastes." UN University, 22 Feb. 2010. Web. 22 Dec. 2015.

[9] Mitch Moxley, "E-Waste Hits China", Inter Press Service, 2011 http://ipsnews.net/news.asp?idnews=56572

[10] Dasgupta, Susmita; Hua Wang; Wheeler, David; (1997-11-30). "Surviving success: policy reform and the future of industrial pollution in China, Volume 1". The World Bank. Retrieved 2009-03-14.

[11] Kahn, Joseph; Jim Yardley; (August 26, 2007). "As China Roars, Pollution Reaches Deadly Extremes". The New York Times. Retrieved 2009-03-14.

[12] "China 'buried smog death finding'". BBC. 2007-07-03.

[13] Wassermann, Rogerio (2009-04-02). "Can China be green by 2020?". BBC.

[14] Vennemo, H.; Aunan, K.; Lindhjem, H.; Seip, H. M. (2009). "Environmental Pollution in China: Status and Trends". *Review of Environmental Economics and Policy* **3** (2): 209. doi:10.1093/reep/rep009.

[15] Kumar, Prashant; Morawska, Lidia; Martani, Claudio; Biskos, George; Neophytou, Marina; Di Sabatino, Silvana; Bell, Margaret; Norford, Leslie; Britter, Rex (2015-02-01). "The rise of low-cost sensing for managing air pollution in cities". *Environment International* **75**: 199–205. doi:10.1016/j.envint.2014.11.019.

[16] "PM2.5". *United States Environmental Protection Agency*. Retrieved 7 October 2014.

[17] T, Phillips (5 November 2013). "China's air pollution blamed for eight-year-old's lung cancer". The Telegraph. Retrieved 7 October 2014.

[18] Jonathan Watts, Air pollution could become China's biggest health threat, expert warns, Friday 16 March, The Guardian, http://www.guardian.co.uk/environment/2012/mar/16/air-pollution-biggest-threat-china

[19] Wong, Edward (April 3, 2013). "2 Major Air Pollutants Increase in Beijing". *The New York Times*. Retrieved April 4, 2013.

[20] Kaiman, Jonathan (16 February 2013). "Chinese struggle through 'airpocalypse' smog". *The Guardian* (London). Retrieved 4 March 2013.

[21] Lu, Z.; Streets, D. G.; Zhang, Q.; Wang, S.; Carmichael, G. R.; Cheng, Y. F.; Wei, C.; Chin, M.; Diehl, T.; Tan, Q. (2010). "Sulfur dioxide emissions in China and sulfur trends in East Asia since 2000". *Atmospheric Chemistry and Physics* **10** (13): 6311. doi:10.5194/acp-10-6311-2010.

[22] "Pollution makes cancer the top killer". Xie Chuanjiao (China Daily). 2007-05-21.

[23] "2007 World Development Indicators: Air Pollution." Table 3.13.. World Bank (2007). Washington, DC.

[24] Hennock, Mary (1 March 2012). "China combats air pollution with tough monitoring rules". *The Guardian*.

[25] "US Embassy Accidentally Calls Beijing's Pollution 'Crazy Bad'". Techdirt. 2010-11-23. Retrieved 2013-01-21.

[26] "Beijing's polluted air defies standard measure". Ctv.ca. 2011-02-26. Retrieved 2013-01-21.

[27] Barbara Demick (2011-10-29). "U.S. Embassy air quality data undercut China's own assessments". Los Angeles Times. Retrieved 2013-01-21.

[28] "Pollution in Beijing Reach Beyond Index Levels". 2nd-greenrevolution.com. 2011-12-13. Retrieved 2013-01-21.

[29] "Foreign embassies' air data issuing inaccurate, unlawful: official". Xinhua, 5 June 2012

[30] BNO News (5 June 2012). "China asks U.S. Embassy to stop publishing Beijing air quality data", Channel 6 News.

[31] "Beijing, China Air Pollution Hits Hazardous Levels". Huffingtonpost.com. 2013-01-12. Retrieved 2013-01-21.

[32] "BBC News - Beijing air pollution soars to hazard level". Bbc.co.uk. 2013-01-12. Retrieved 2013-01-21.

[33] "China: record smog levels shut down city of Harbin | euronews, world news". Euronews.com. Retrieved 2013-10-21.

[34] Andrews-Speed, Philip (November 2014). "China's Energy Policymaking Processes and Their Consequences". *The National Bureau of Asian Research Energy Security Report*. Retrieved December 24, 2014.

[35] Usman W. Chohan (May 2014). "An Eco-friendly Exodus: Heavy Industry in Beijing 环保政策". *McGill University Economic Publications*.

[36] Stanway, D (6 November 2013). "China cuts gas supply to industry as shortages hit". Reuters.

[37] Boren, Zachary Davies (27 August 2015). "China air pollution: Beijing records its cleanest air ever". Retrieved 29 August 2015.

[38] "China to Unify Color-coded Pollution Alert System". China Radio International. 2014-12-03.

[39] "Beijing raises 'red alert' threshold for air pollution warning". The Guardian. 2016-02-21.

[40] " 环保部: 京津冀 6 城统一重污染预警分级". 新华网. 2016-02-05.

[41] LaFraniere, Sharon (2011-06-15). "Lead Poisoning in China: The Hidden Scourge". New York Times.

[42] He, K.; Wang, S.; Zhang, J. (2009). "Blood lead levels of children and its trend in China". *Science of the Total Environment* **407** (13): 3986–3993. doi:10.1016/j.scitotenv.2009.03.018. PMID 19395068.

[43] "The People's Republic of China: National Implementation Plan for the Stockholm Convention on Persistent Organic Pollutants" (PDF). Stockholm Convention on Persistent Organic Pollutants. 2007.

[44] "Swimming in Poison: A hazardous chemical cocktail found in Yangtze River Fish". Greenpeace China. 2010-08-26.

[45] Wang Ying. "Operation blitzkrieg against desert storm". *China Daily*. Archived from the original on April 10, 2007. Retrieved April 3, 2007.

[46] "Ill Winds". *Science News Online*. Archived from the original on March 19, 2004. Retrieved October 6, 2001.

[47] "N-hexane Poisoning Scare At Apple Supplier In China". China Tech News. 2010-02-22.

[48] *China to Survey Soil Amid Fears Over Rice* June 12, 2013 Wall Street Journal

[49] Sun Xiaohua (2007) "Call for return to green accounting", "China Daily", 19 Apr 2007.

[50] Cost of Pollution in China Economic Estimates of Physical Costs. 2007. World Bank. http://siteresources.worldbank.org/INTEAPREGTOPENVIRONMENT/Resources/China_Cost_of_Pollution.pdf

[51] Polluting China for the sake of economic growth. 27-Apr-2012. EurekAlert!. http://www.eurekalert.org/pub_releases/2012-04/ip-pcf042712.php

[52] Associated Press (2015-08-13). "Air pollution in China is killing 4,000 people every day, a new study finds". *The Guardian*. Retrieved 2015-08-14.

[53] Hornby, L (22 October 2013). "Cleaner air a bitter pill for north China cities". Financial Times.

[54] Qin, Jize (2004-07-15). "Most polluted cities in China blacklisted". China Daily.

[55] "WEATHER & EXTREME EVENTS 7 of 10 Most Air-Polluted Cities Are in China". *JAN 16, 2013* (Imaginechina/Corbis). http://news.discovery.com. Retrieved 1 September 2014.

4.4.8 External links

- Real-time air quality index map

Articles

- Cleaner Production in China - Current and comprehensive information source on China's campaign to reduce pollution

- Photo essay on water pollution in Huai River Basin

- Most polluted cities in China

- Clearing the Air: China's Environmental Challenge - Asia Society - Overview on China air pollution problem

- Documentary project "Pollution in China."

- *Spill in China Underlines Environmental Concerns* March 2, 2013 The New York Times

Videos

- "The Environmental Challenge to China's Future", Dr. Elizabeth Economy, 2010

- Accompanying the growth of industry is an increase in pollution and toxic waste that threatens the livelihood and health of people in rural fishing and farming communities.

- Youtube video:China'{}s Pollution Busters

- Terrible Pollution in China

- Environmental activist Wu Dengming documents.

- Youtube video:Where does e-waste end up?

- Youtube video:Exporting Harm trailer

- Youtube video:Electronic Trash Village - China

4.5 SDAS UAV

SDAS UAV is a Chinese UAV developed by Shandong Academy of Sciences (SDAS, 山东省科学院) specifically for pollution surveillance missions, and it has entered service with the local governmental environmental establishment, Shandong Environmental Protection Department (山东省环保厅) since June 2014.

4.5.1 Octocopter

The UAV developed by SDAS for environmental monitoring applications is an octocopter, and it is developed by the UAV project development team (无人机项目研发团队) of Shandong Computer Science Center (山东省计算中心) of SDAS, and the team is led by Dr. Liu Xiang (刘祥).*[1] The need for aerial pollution monitoring system was urgent because local businesses were not (and still are) not honest about the pollution control practice: although every local business is forced to install Continuous Emission Monitoring System (CEMS, 烟气连续监测系统), they could and often turn off the system when release smog or other pollutants, and even destroy data in order to lower the amount emission to meet the quota, by simply claiming the equipment had malfunctioned. It was very difficult to catch such illegal acts because every CEMS is on site and there was simply not enough manpower to monitor every system.*[2] An alternative would be monitor the pollution emission from air, where businesses could not tamper with monitoring system, hence the UAV for environmental monitoring was born.*[3] Since 2012, SDAS had joined forces with Australian National University (ANU) to develop an UAV for pollution monitoring, with ANU providing assistance to develop the onboard monitoring system, and Chinese are responsible for the aerial platform itself.*[2] The resulting UAV is an octocopter with a pair of skids as landing gear, and only requires three to five square meters to take-off and land, and capable of monitor pollutants of size as small as PM 2.5.*[4]*[1]*[2] Specification:*[1]*[3]

- Size (m): 1.1

- Weight (kg): < 10

- Normal operating altitude (m): 500

4.5.2 See also

List of unmanned aerial vehicles of the People's Republic of China

4.5.3 References

[1] "SDAS Octocopter". Retrieved May 17, 2014.

[2] "SDAS Octorotor". Retrieved Apr 9, 2014.

[3] "SDAS UAV". Retrieved May 26, 2014.

[4] "SDAS unmanned aerial vehicle". Retrieved May 17, 2014.

4.6 Yunnan Jinding Zinc

Yunnan Jinding Zinc Corporation Ltd., founded in 2003, is a subsidiary of Sichuan Hongda, a larger mining firm which itself is a subsidiary of the Hanlong Group. It operates a mine and smelter near Jinding in Lanping County, Yunnan with an annual capacity of over 100,000 tons of zinc.*[1]

4.6.1 Pollution

The firms smelter operations are near residential housing in Jinding and other villages which has resulted in high levels of lead, zinc, and cadmium in the soil surrounding the plant and high levels of lead in the blood of residents. Environmental enforcement efforts by the Chinese and Yunnan governments have been ineffective.*[2] High levels of lead in household dust was found in residences in the area. Lead in household dust is one way lead is ingested by children.*[3]

4.6.2 References

[1] "Company Overview of Yunnan Jinding Zinc Co., Ltd". *Businessweek* (Bloomberg). Retrieved June 12, 2015.

[2] "Greenpeace: Investigation finds pollution and illness ignored at Asia's largest lead mine, Yunnan Province" (Press release). *greenpeace.org*. Greenpeace China. June 9, 2015. Retrieved June 12, 2015.

[3] Cherie Chan (June 12, 2015). "Soil Contamination Found Near Huge Mine in Western China". *The New York Times*. Retrieved June 12, 2015.

4.6.3 External links

- "360 drone investigation: Poisoned towns" Greenpeace YouTube video of tailings ponds and smelter

Chapter 5

Waste in China

5.1 EcoPark (Hong Kong)

EcoPark (Chinese: 環保園) located in Tuen Mun Area 38, on west side of Hong Kong, is similar to an industrial park exclusively for waste recycling and environmental engineering. This is the first of its kind in Hong Kong.

5.1.1 Introduction

In December 2003, the Hong Kong Government mapped out a strategy on waste management emphasized waste reduction and recovery. Hong Kong currently recycles 48% of its municipal solid waste (MSW), but over 99% of recovered recyclable materials are exported to Mainland China for further re-processing while less than 1% are treated locally and re-manufactured into useful products. With the measures to promote waste recovery, recycling and reuse in place, a local waste management area like EcoPark is a viable option for furthering Hong Kong's recycling program.

EcoPark aims to promote the local recycling industry and jump-start a circular economy to provide a sustainable solution to the city's waste problems. By encouraging and promoting the reuse, recovery and recycling of waste resources and returning them to the consumption loop, the EcoPark will help realize the full potential of the local recycling industry and alleviate the heavy reliance on the export of recyclable materials recovered from Hong Kong.

5.1.2 Design and construction

The EcoPark occupies 200,000 square metres of land in Tuen Mun Area 38 and will be developed in two phases. As pledged in the Policy Framework, the aim is to commission Phase I of EcoPark (80,000 square metres) towards the end of 2006 and Phase II (120,000 square metres) in 2009. Hong Kong Government funding will be used to build the basic infrastructure of EcoPark.

The EcoPark will be divided into lots of different sizes.

Lots in EcoPark will be tendered for specific recovered materials and processes that help achieve Hong Kong's government waste management objectives, in particular, in recycling local wastes. Admission criteria will be developed with priority given to processes involving value-added technologies, and target materials of the proposed Producer Responsibility Schemes.

5.1.3 Progress

All six lots in EcoPark Phase I has been allocated for recycling of waste cooking oil, waste computer equipment, waste metals, waste wood, waste plastics and waste car batteries. Some of them are already in operation, while the rest of tenants will start their operation shortly.

Construction works of the EcoPark Phase II have already completed. Two lots have been allocated to non-government organisations for recycling of plastics and waste electrical appliances. The remaining lots in EcoPark Phase II will be available for tendering in late 2010.

5.1.4 Phase I tenants

Champway Technology Limited - Recycling of waste cooking oil into biodiesel
Li Tong Group - Recycling of waste computer equipment
Shiu Wing Steel Limited - Recycling of waste metals
Hong Kong Hung Wai Wooden Board Company - Recycling of waste wood
Hong Kong Telford Envirotech Company Limited - Recycling of waste plastics
Cosmos Star Company Limited - Recycling of car batteries

5.1.5 Phase II lots

Tendering of Phase II lots (with a total area of 100,000 square metres) will start in late 2010. Further information on the tendering can be obtained from EcoPark management office.

5.1.6 Visitor Centre

The 1,000-square metre EcoPark Visitor Centre is the first education centre in Hong Kong with a main theme of solid waste management. Admission is free and docent service will be provided. Online and telephone booking can be arranged through the EcoPark Management Office.

5.1.7 Public consultation

Hong Kong Government has consulted the Tuen Mun District Council and members support the development of Eco-Park and agree that EcoPark will help promote development of local recycling industry and create job opportunities in Tuen Mun. The Council hopes that EcoPark will become a landmark for Tuen Mun.

Local trade associations and recyclers were also consulted and they support the development of EcoPark, agreeing that by providing long-term land at affordable cost, together with supporting infrastructure, EcoPark will help enhance recycling technology development and improve waste recovery rates in Hong Kong.

5.1.8 Environmental considerations

An Environmental Impact Assessment (EIA) was carried out in respect of air and water quality, waste management, land contamination, landfill gas hazard and hazard to life, in which a wide range of recycling processes for different material types were examined. The assessment recommends a list of materials and processes to be allowed and also recommends a number of mitigation measures. With these measures in place, the EIA concludes that there will be no significant environmental impacts to the surrounding areas.

5.1.9 External links

- Official website

- EcoPark Management Office

- EcoPark's Administration Building wins the BCI Asia Green Design Award

Coordinates: 22°22′00.2″N   113°55′38.01″E   /
22.366722°N 113.9272250°E

5.2 Electronic waste in China

Electronic waste in China is a serious environmental issue. The amount of electronic waste (e-waste) is increasing due to rising economies like China and India and a higher demand of electronic devices combined with a shorter economic lifespan in the Western world.[1] Though e-waste from the Western world is responsible for a large portion of the e-waste, the biggest threat comes from other regions in the world like India, Thailand, and China itself.[2] Roughly 70% of global e-waste ends up in China.[3] As a result, China has to deal with the environmental damage and health problems related to the increasing amount of e-waste. Most of these problems arise from the fact that 60% of the e-waste is processed in informal recycling centres by unskilled ill-equipped manual labour.[4]

5.2.1 E-waste

China in 2011 was the world's second largest producer of electronic waste and produced 2.3 million tones. The amount is expected to increase as the Chinese economy grows. Large amounts of foreign electronic waste are also imported. Disposal of electronic waste can create jobs and recycle valuable metals but also harm humans and the environment by releasing pollutants. Legislation banning importation and requiring proper disposal of indigenous waste as well as a governmental subsides for proper disposal have recently been introduced but have been criticized as insufficient and susceptible to fraud. There have been local successes, such as the city of Tianjin where 38,000 tonnes were disposed of properly in 2010, but much electronic waste is improperly handled.[5]

China receives pollution from both ends of the supply chain: during production process and by allowing electronic waste to be recycled and dumped in the country.[6]

5.2.2 Affected regions

Main article: Electronic waste in Guiyu

The main region where the e-waste is shipped to is the Guangdong province, situated along China's south east coast. From there it is spreading to other regions such as Zhejiang, Shanghai, Tianjin, Hunan, Fujian and Shandong. All of these regions are located along China's entire east coast.[2] Guiyu in Guangdong Province is the location of the largest electronic waste site on earth.[7]

5.2.3 Process of e-waste

In China there is a formal and a strong informal collecting system. The informal collection system called "cherry picking" utilizes only recyclable appliances and sells the reusable pieces to the local second-hand market. Concerning the formal sector there are some collecting projects put in place.[8] Furthermore, the process of electric and electronic devices contains many toxic substances and because the processing of e-waste is being done by burning and heating to extract valuable material the health hazards are omnipresent. The toxics that are released during this process are very harmful to a person's health. Given that most of the recycling is done improperly and without the necessary safety precautions, e-waste is directly responsible for deteriorating health and environment in China's east coast.[9]

5.2.4 Informal sector

Informal recycling generally uses primitive processes and does not have the appropriate facilities to safeguard environmental and human health.[10] Nonetheless, it is a very profitable market in China thanks to low wages, high demand for used electronics, used parts and materials.[11] The informal recycling method consists mainly of manual, unskilled labor and is inherently mobile. Therefore, regulations might not be as effective as intended. Moreover, this industry "feeds" thousands of families.[9]

5.2.5 Solutions

Attempts to control the informal sector

In the regions of Tianjin, Taicing, Ningbo, Taizhou and Zhangzhou, local recycling parks have been built in which informal laborers still work as manual recyclers, but then under production and pollution management.[12] In Guiyu, a different solution was found. Here, the government promoted technical upgrade in the informal workshops by replacing coal-fired grills with electrical heaters when taking out components from circuit boards.[9]

Basel Convention

The problem has to be tackled top-down by government and UN-based regulations, like the Basel Convention, to control the processing and transporting of e-waste and to provide environmental and social justice.[13] The United Nations (UN) Basel Convention on the Control of Transboundary Movements of Hazardous Wastes and Their Disposal is the most far-reaching regulation that exists on a global scale to

address e-waste. However, the lucrative business that is created by e-waste recycling is responsible for the undermining of this convention in areas where e-waste is transported to.[14]

Chinese legislation

The Chinese government has taken actions as well. Initially there was a complete ban on improper recycling but this was quickly dropped. They have now issued a variety of environmental laws, regulations, standards, technical guidance and norms related to electronic product production and e-waste management.[11] Nevertheless, laws and regulations put in place by the Chinese government lack of adequate resources to enforce them. Moreover, the financial windfall associated with e-waste makes these laws and regulations weak.[13] In 2008, The Chinese State Council also approved a "draft regulation on the management of electronic waste." This regulation is intended to promote the continued use of resources through recycling and to monitor the end-of-life treatment of electronics. Under the new regulations, recycling of electronics by the consumer is mandated. It also requires the recycling of unnecessary materials discarded in the manufacturing process.[15] The Management Regulations for Recycling and Disposing of Consumer Electronics and Electronic Waste, intended to be effective January 1, 2011, bans import of toxic e-waste, requires treatment of e-wastes to have license, and treatment plants to treat pollution.[16] One of the most successful policies is probably the Extended Producer Responsibility (EPR). EPR makes manufacturers responsible for electronics collection and recycling. Therefore, the producer is more involved in the life cycle of a product.[11]

Provincial Chinese programs

There are different examples in the region of Qingdao, Beijing and the Sichuan provinces, where the current projects are developed.[8]

A big issue can be found in the Sichuan Province, close to the Tibetan border, where people had a habit of throwing waste in rivers and nature. Local leaders, among others monks and village representatives, decided to call for help from Norlha to design a region specific program. Monks have been informed about the proper way to dispose of (e-)waste, which they could pass through in religious celebrations. At the same time posters have been handed out to the communities and children have been informed by the NGO in their schools. Moreover, in 5 villages waste collections systems and storage points for e-waste have been created.[17]

Another project is the "Home Appliance Old for New Rebate Program", which was first launched in nine cities and provinces who are considered as economically developed regions. It is a recycling system, where only accredited collectors who usually work in the retail industry can collect and take back old appliances from consumers and reward these actions with discount coupons. Since only authorized collectors were participating in the process, it gives the possibility to pay the consumers a higher price for their e-waste[8]

Corporate initiatives

Many companies, like Nintendo, are aware of the problem of e-waste and are developing their own initiatives. Companies joined forces by creating a collective e-waste reclamation campaign. But that does not solve the whole problem.[11]

In response to low incentives some companies, like Dell, started to provide compensations to consumers in Beijing and Shanghai of US$0.15 for 1 kg of old computer. In order to receive the incentive consumers had to bring their used computers to local Dell stores at their own expense. The project failed because the financial gains of returning their computer to formal recyclers were lower than the gains from selling computers to informal collectors.[11]

5.2.6 Legislative inadequacies

Even though legislation and regulations have been accepted by the developed countries against illegal exportation of e-waste, the high number of illegal shipments is contributing to the bad situation of e-waste in China.[18] For instance, the members of the EU agreed not to transport any waste subject to the Basel Convention out of the EU or the OECD but illegal shipments are still rising in China and other developing countries.[19] Greenpeace International claims that a large amount of e-waste is usually illegally shipped from Europe, the U.S. and Japan to China. One of the main incentives for them to export e-waste is that the cost of domestic e-waste disposal is higher than the exportation fees.[20] Moreover, e-waste brokers make large profits from the trade and get paid twice: once for acquiring the e-waste, once for shipping it.[13]

In China, informal collectors buy old electronic devices from consumers. The incentive to participate in collection systems, which cost them compared to informal recycling, is low, even though many Chinese consumers realize that it is important to recycle e-waste safely.[11] As many as 90% of the consumers are reluctant to pay for e-waste recycling because there is still monetary value in the end-life of products.[2]

5.2.7 See also

- Electronic waste by country
- Environmental issues in China
- China RoHS

5.2.8 References

[1] Robinson, Brett H. (2009). "E-waste: An assessment of global production and environmental impacts". *Science of the Total Environment* **408** (2): 183–91. doi:10.1016/j.scitotenv.2009.09.044. PMID 19846207.

[2] Liu, Xianbing; Tanaka, Masaru; Matsui, Yasuhiro (2006). "Electrical and electronic waste management in China: Progress and the barriers to overcome". *Waste Management & Research* **24** (1): 92–101. doi:10.1177/0734242X06062499. PMID 16496875.

[3] "70% of annual global e-waste dumped in China". CRI. May 24, 2012.

[4] Martin, Eugster; Hongjun, Fu (August 18, 2004). *e-Waste Assessment in P.R. China* (PDF). Swiss Federal Laboratories for Materials Science and Technology.

[5] Moxley, Mitch (July 21, 2011). "E-Waste Hits China". Inter Press Service.

[6] "Dirty Secrets". ABC News. October 26, 2010.

[7] Johnson, Tim (April 9, 2006). "E-waste dump of the world". *The Seattle Times*. Knight Ridder Newspapers.

[8] Wang, Feng; Kuehr, Ruediger; Ahlquist, Daniel; Li, Jinhui (April 5, 2013). *E-waste in China: A country report* (PDF). United Nations University Institute for Sustainability and Peace.

[9] Chi, Xinwen; Streicher-Porte, Martin; Wang, Mark Y.L.; Reuter, Markus A. (2011). "Informal electronic waste recycling: A sector review with special focus on China". *Waste Management* **31** (4): 731–42. doi:10.1016/j.wasman.2010.11.006. PMID 21147524.

[10] Green Peace (March 2004). "Key findings from Taizhou Field Investigation" (PDF). Basel Action Network.

[11] Yu, Jinglei; Williams, Eric; Ju, Meiting; Shao, Chaofeng (2010). "Managing e-waste in China: Policies, pilot projects and alternative approaches". *Resources, Conservation and Recycling* **54** (11): 991–9. doi:10.1016/j.resconrec.2010.02.006.

[12] Shinkuma, Takayoshi; Nguyen Thi Minh Huong (2009). "The flow of E-waste material in the Asian region and a reconsideration of international trade policies on E-waste". *Environmental Impact Assessment Review* **29**: 25–31. doi:10.1016/j.eiar.2008.04.004.

[13] Puckett, Jim; Byster, Leslie; Westervelt, Sarah; Gutierrez, Richard; Davis, Sheila; Hussain, Asma; Dutta, Madhumitta (February 25, 2002). Puckett, Jim; Smith, Ted, eds. *Exporting Harm: The High-Tech Trashing of Asia* (PDF). The Basel Action Network & Silicon Valley Toxics Coalition.

[14] Schluep, Mathias; Hagelueken, Christian; Kuehr, Ruediger; Magalini, Federico; Maurer, Claudia; Meskers, Christina; Mueller, Esther; Wang, Feng (July 2009). Sonnemann, Guido; de Leeuw, Bas, eds. *Recycling – From E-Waste To Resources* (PDF). Sustainable Innovation and Technology Transfer Industrial Sector Studies. United Nations Environment Programme.

[15] Hoggard, Stuart (August 28, 2008). "China approves e-waste regulation – systems proposed, penalties established". PackWebAsia. Archived from the original on November 21, 2008.

[16] Chen, Chu (October 14, 2010). "Point of View: ELAW's Intern looks at China's e-waste industry". Environmental Law Alliance Worldwide.

[17] "Project in progress: Waste management in China". Norlha. 2012.

[18] Ni, Hong-Gang; Zeng, Eddy Y. (2009). "Law Enforcement and Global Collaboration are the Keys to Containing E-Waste Tsunami in China". *Environmental Science & Technology* **43** (11): 3991–4. doi:10.1021/es802725m. PMID 19569320.

[19] *Waste without borders in the EU? Transboundary shipments of waste.* European Environment Agency. 2009. doi:10.2800/14850. ISBN 978-92-9167-986-7.

[20] Williams, Eric; Kahhat, Ramzy; Allenby, Braden; Kavazanjian, Edward; Kim, Junbeum; Xu, Ming (2008). "Environmental, Social, and Economic Implications of Global Reuse and Recycling of Personal Computers". *Environmental Science & Technology* **42** (17): 6446–54. doi:10.1021/es702255z. PMID 18800513.

5.3 Electronic waste in Guiyu

Guiyu, in Guangdong Province, China, is an agglomerate of four adjoined villages widely perceived as the largest electronic waste (e-waste) site in the world.[1] In 2005 there were 60,000 e-waste workers in Guiyu who processed the more than 100 truckloads that were transported to the 52 square kilometre area every day.[2] The constant movement into and processing of e-wastes in the area leading to the harmful and toxic environment and living conditions, coupled with inadequate facilities, have led to the Guiyu town being nicknamed the "electronic graveyard of the world".[3]

E-waste pile at Guiyu.

5.3.1 Health impacts

Many of the primitive recycling operations in Guiyu are toxic and dangerous to workers' health with 80% of children suffering from lead poisoning.[4] Above-average miscarriage rates are also reported in the region. Workers use their bare hands to crack open electronics to strip away any parts that can be reused—including chips and valuable metals, such as gold, silver, etc. Workers also "cook" circuit boards to remove chips and solders, burn wires and other plastics to liberate metals such as copper; use highly corrosive and dangerous acid baths along the riverbanks to extract gold from the microchips; and sweep printer toner out of cartridges. Children are exposed to the dioxin-laden ash as the smoke billows around Guiyu, and finally settles on the area. The soil has been saturated with lead, chromium, tin, and other heavy metals. Discarded electronics lie in pools of toxins that leach into the groundwater, making the water undrinkable to the extent that water must be trucked in from elsewhere. Lead levels in the river sediment are double European safety levels, according to the Basel Action Network.[5] Lead in the blood of Guiyu's children is 54% higher on average than that of children in the nearby town of Chendian.[6] Piles of ash and plastic waste sit on the ground beside rice paddies and dikes holding in the Lianjiang river.

A recent study of the area evaluated the extent of heavy metal contamination from the site. Using dust samples, scientists analysed mean heavy metal concentrations in a Guiyu workshop and found that lead and copper were 371 and 115 times higher, respectively, compared to areas located 30 kilometres away.[7] The same study revealed that sediment from the nearby Lianjiang River was found to be contaminated by polychlorinated byphenyls at a level three times greater than the guideline amount.

5.3.2   Economic rationale

The economic incentives created by strict domestic regulation, non-existent or unenforced regulations in developing countries, and the ease of free trade brought about by globalization, led recyclers to export e-waste. The value of parts in discarded electronics provides an incentive for poverty-stricken citizens to migrate to Guiyu from other provinces to work in processing it. The average worker, adult or child, makes barely $1.50/day (or 17 cents/hour). The average workday is sixteen hours. This $1.50 is made by recovering the valuable metals and parts that are within the piles of discarded electronics. Even this relatively tiny profit is enough motivation for workers to risk their health.*[8]

5.3.3   Agriculture

Once a rice village,*[9] the pollution has made Guiyu unable to produce crops for food and the water of the river is undrinkable.

5.3.4   Media coverage

Guiyu as an e-waste hub was first documented fully in December 2001 by the Basel Action Network, a non-profit organization which combats the practice of toxic waste export to developing countries in their report and documentary film entitled *Exporting Harm*.*[5] The health and environmental issues exposed by this report and subsequent scientific studies*[10] have greatly concerned international organisations such as the Basel Action Network and later Greenpeace and the United Nations Environment Programme and the Basel Convention. Media documentation of Guiyu is tightly regulated by the Chinese government, for fear of exposure or legal action. For example, a November 2008 news story by *60 Minutes*, a popular US TV news program, documented the illegal shipments of electronic waste from recyclers in the US to Guiyu. While taping part of the story on-site at an illegal recycling dump in Guiyu, representatives of the Chinese recyclers attempted without success to confiscate the footage from the *60 Minutes* TV crew.*[11] Greenpeace has protested the environmental impacts of e-waste recycling in Guiyu using different methods to raise awareness such as building a statue using e-waste collected from a site in Guiyu, or delivering a truckload of e-waste dumped in Guiyu back to Hewlett Packard headquarters. Greenpeace has been lobbying large consumer electronics companies to stop using toxic substances in their products, with varying degrees of effectiveness.*[12]

5.3.5   Cleanup efforts

Since 2007, conditions in Guiyu have changed little despite the efforts of the central government to crack down and enforce the long-standing e-waste import ban. Recent studies have revealed some of the highest levels of dioxin ever recorded. However, because of the work of activist groups and increasing awareness of the situation, there is hope for the site to be improved. "It can be done. Look at what happened with lead acid batteries. We discovered they were hazardous, new legislation enforced new ways of dealing with the batteries which led to an infrastructure being created. The key was making it easy for people and companies to participate. It took years to build. E-waste is going the same route. But attitudes have changed and we will get there," says Robert Houghton, president and founder of Redemtech, an asset management and recovery firm.*[13] Zheng Songming, head of the Guiyu Township government has published a decree to ban burning electronics in fires and soaking them in sulfuric acid, and promises supervision and fines for violations. Over 800 coal-burning furnaces have been destroyed because of this ordinance, and most notably, air quality has returned to Level II, now technically acceptable for habitation.

5.3.6   See also

- Electronic waste in China

- Environmental issues in China

- Pollution in China

5.3.7   References

[1] Johnson, Tim (April 9, 2006). "E-waste dump of the world" . The Seattle Times. Retrieved 2007-03-09.

[2] "China focus: Chinese recycling base in pursuit of sustainable development" . Xinhua General News Service. May 23, 2005.

[3] Yeung, Miranda (April 21, 2008). "There's a dark side to the digital age" . South China Morning Post (Guangdong, China).

[4] Monbiot, George (September 21, 2009). "From toxic waste to toxic assets, the same people always get dumped on" . The Guardian (London).

[5] "Exporting Harm: The High-Tech Trashing of Asia" (pdf). Basel Action Network. February 25, 2002.

[6] http://ehp.niehs.nih.gov/realfiles/members/2007/9697/9697.html

[7] Leung, Anna (March 4, 2008). "Heavy Metals Concentrations of Surface Dust from e-Waste Recycling". Hong Kong.

[8] "Waste not want not? Not in the world of computers". *Business Daily Update*. September 27, 2006.(registration required)

[9] "You'll never think the same way again". July 2010.

[10] "Scientific Articles". Basel Action Network.

[11] "Following The Trail Of Toxic E-Waste". *cbsnews.com*. CBS News. Retrieved 26 March 2015.

[12] Chi-Chu, Tschang (May 24, 2005). "Greenpeace launches e-waste drive in China". *The Straits Times* (Singapore).

[13] "'Clean-tech' start-ups are pushing the green button". United Nations Environment Programme.

5.3.8 External links

- Viceland: CTRL+ALT+LANDFILL – China's Secret Computer Graveyard

- The complete photo service from Vice's photographer Luca Gabino

- Video about e-waste in Guiyu

- Series of pictures of e-waste in Guiyu -- Greenpeace China

- FOXNews.com - Chinese Recyclers Live in Toxic E-Waste Dump - Science News | Science & Technology | Technology News

- Photo report, China's electronic waste village, TIME magazine

5.4 Food and Environmental Hygiene Department

Food and Environmental Hygiene Department (Chinese: 食物環境衞生署), or **FEHD** (食環署) for short, is a department of Hong Kong Government, reporting to the Health, Welfare and Food Bureau. It is responsible for food hygiene and environmental hygiene. It replaced part of the role of the Urban Council and the Urban Services Department, and the Regional Council and the Regional Services Department.

5.4.1 History

The department was founded in 2000 along with the Leisure and Cultural Services Department. These two departments were intended to absorb most of the functions of the former Urban Council and Regional Council, which were dissolved at that time.

The Chief Executive announced in October 2005 in his policy address the plan to restructure the department into a new Food Safety, Inspection and Quarantine Department, with the responsibility of environmental hygiene transferred to the Agriculture, Fisheries and Conservation Department.

5.4.2 Duties and operations

The department's responsibilities include food inspection, food standards, food safety licence issuing, refuse collection, street washing, hawker control, operation of Lunar New Year fairs, and municipal building cleansing. Facilities operated by the department include markets, cooked food centres, public toilets, slaughterhouses, refuse collection points, cemeteries, crematoria, and columbaria.

Health Education Exhibition and Resources Centre

It now manages the Health Education Exhibition and Resource Centre in Kowloon Park.

5.4.3 Directors

- Rita Lau (2000–2002)*[2]

- Gregory Leung Wing-lup (2002–2006)

- Eddy Chan Yuk-tak (2006–2007)*[3]

- Warner Cheuk Wing-hing (2007–2010)

- Clement Leung (2010–2014)

- Vivian Lau Lee-kwan (2014–present)

5.4.4 See also

- 2016 Mong Kok civil unrest

5.4.5 References

[1] "Estimates for the year ending 31 March 2015" (PDF). *2014-15 Budget*. Hong Kong Government. Retrieved 14 November 2014.

[2] "Appointment of new Chairman of Public Service Commission". Hong Kong Government. 10 April 2014. Retrieved 14 November 2014.

[3] "Senior Appointments". Hong Kong Government. 8 November 2007. Retrieved 10 November 2014.

5.4.6 External links

- Official website

5.5 Gin Drinkers Bay

Gin Drinkers Bay (Chinese: 醉酒灣; literally: "Drunkard's Bay") or **Gin Drinker's Bay**, also known as **Lap Sap Wan** (Chinese: 垃圾灣; Jyutping: *laap6 saap3 waan1*; literally: "Rubbish Bay"), was a bay in Kwai Chung, Hong Kong.

The bay was reclaimed in 1960s and became Kwai Fong and part of Kwai Hing. At the mouth of the bay stood the island of Tsing Chau.

The bay was a harbour for Tanka fishing junks. They relocated to Tsing Yi Tong and Mun Tsai Tong of Tsing Yi Island before the commencement of reclamation.

Lap Sap (垃圾) means "rubbish" in Cantonese. It is unclear that why the bay was named 'rubbish' in the past. However, coincidentally it was once a dumping area for rubbish after extensive reclamation.[1] It is assumed that in Gin Drinkers Bay Park or Kwai Chung Park near Tsing Chau that the area is subject to landfill gas produced deep in the ground even though it is covered with hills of earth. It remains closed due to unsafe levels of landfill gas.

Gin Drinkers Bay is known for the Gin Drinkers Line, which formed a defensive line against the Japanese invasion in 1941.

5.5.1 See also

- Kwai Chung Incineration Plant

- Waste management in Hong Kong

5.5.2 References

[1] Bray, Denis (2001). *Hong Kong Metamorphosis*. Aberdeen, Hong Kong: Hong Kong University Press. pp. 80–81.

Coordinates: 22°21′21″N 114°06′53″E / 22.35573°N 114.11460°E

5.6 Inventory of hazardous materials

An **Inventory of Hazardous Materials** is one of the requirements of the Hong Kong convention for the 'safe and environmentally sound recycling of ships' (Hong Kong Convention) which was adopted in May 2009.[1] The conference that created the convention was attended in 2009 by members of 63 countries, and overseen by the International Maritime Organisation (IMO), which is a specialist agency of the United Nations (U.N).[2]

The Hong Kong Convention has been designed to try to improve the health and safety of current ship breaking practices. At present, this is a very dangerous practice, whereby large ships are beached and then dismantled by hand by workers with very little personal protective equipment (PPE). This is most common in Asia, with India, Bangladesh, China, and Pakistan holding the largest ship breaking yards.[3]

The Hong Kong Convention recognised that ship recycling is the most environmentally sound way to dispose of a ship at the end of its life, as most of the ship's materials can be reused. However, it sees current methods as unacceptable. The work sees many injuries and fatalities to workers, as they lack the correct safety equipment to handle the large ship correctly as it is dismantled and most vessels contain a large amount of hazardous materials such as asbestos, PCBs, TBT, and CFCs, which can also lead to highly life-threatening diseases such as mesothelioma and lung cancer.[4]

The Inventory of Hazardous Materials has been designed to try to minimise the dangers of these hazards. The Convention defines a hazard as: "any material or substance which is liable to create hazards to human health and/or the environment".[5]

All vessels over 500 gross tonnes (GT) that are in commercial service (the convention does not apply to warships or naval auxiliary or ships operating their whole life only in waters subject to the sovereignty or jurisdiction of the State whose flag the ship is entitled to fly) will have to comply with the convention once it comes into force. Each party that does wish to comply must restrict the use of hazardous materials on all ships that fly the flag of that party.[6]

New ships must all carry an Inventory of Hazardous Materials. The inventory will list all 'hazardous materials' on board the vessel, including their amounts and locations. Existing ships must comply no later than five years after the convention comes into force, or prior to being recycled if this occurs before the five-year period. The inventory will remain with a vessel throughout its lifespan, being updated as all new installations enter the ship, as these may potentially contain hazards. The presence of the inventory will then ensure the safety of crew members during the vessel's working life, and also the safety of workers during the recycling process.

The convention held a fixed deadline for states to sign up between 1 September 2009 and 31 August 2010, and after this remains open for accession. It will enter into force two years after "15 states, representing 40% of the world merchant shipping by gross tonnage, and on average 3% of recycling tonnage for the previous 10 years, have either signed it without reservation as to ratification, acceptance or approval, or have deposited instruments of ratification, acceptance, approval or accession with the Secretary General" .*[7]

The EU Ship Recycling Regulation *[8] entered into force on 30 December 2013. Although this regulation closely follows the Hong Kong convention, there are important differences. The Regulation sets out a number of requirements for European ships, European ship owners, ship recycling facilities willing to recycle European ships, and the relevant competent authorities or administrations. It also requires the Commission to adopt a number of acts implementing the Regulation (in particular the European List of ship recycling facilities authorized to recycle ships flying the Union flag). For the Inventory of Hazardous Materials required by the EU regulation, there are additional substances listed as prohibited.*[9]

5.6.1 References

[1] International Maritime Organisation http://ec.europa.eu/environment/waste/ships/pdf/Convention.pdf 'agenda item 8' Retrieved on 22 September 2010

[2] http://www.imo.org/

[3] Mikelis, Nikos.http://www.imo.org/includes/blastDataOnly.asp/data_id%3D23449/shiprecycling.pdf "A statistical overview of ship recycling", September 2007. Retrieved on 22 September 2010

[4] http://www.cancer.gov/cancertopics/factsheet/Sites-Types/mesothelioma Retrieved on 22 September 2010

[5] http://ec.europa.eu/environment/waste/ships/pdf/Convention.pdf Retrieved on 22 September 2010

[6] International Maritime Organisation http://ec.europa.eu/environment/waste/ships/pdf/Convention.pdf 'agenda item 8' Retrieved on 11 March 2015

[7] http://www.imo.org/Environment/mainframe.asp?topic_id=818 Retrieved on 22 September 2010

[8] http://ec.europa.eu/environment/waste/ships/

[9] http://eur-lex.europa.eu/legal-content/EN/TXT/?uri=CELEX:52012PC0118

5.7 Kwai Chung Incineration Plant

Kwai Chung Incineration Plant, with Rambler Channel Bridge at front

Rambler Channel, Kwai Chung Incineration Plant (centre) with Hong Kong Island in the background.

Kwai Chung Incineration Plant (Chinese: 葵涌焚化爐)

was one of four incineration plants in Hong Kong. The plant was built on a 1.4 hectares (3.5 acres) of reclaimed land along Gin Drinkers Bay, Kwai Chung, near Tsing Chau and the Rambler Channel.

The plant was opened in 1978 to process solid waste from Hong Kong to reduce the need to put waste into landfills.

5.7.1   Cessation of operation

In 1989, the Hong Kong Government issued a white paper, *Pollution in Hong Kong - A Time to Act*. After considering the effects of air pollution on the environment and public health, it was decided to cease using incineration to dispose of solid waste. This decision was later suspended and, as of 2008, the Hong Kong Government is considering constructing new incinerators.[1]

In May 1997, the Kwai Chung Incineration Plant ceased to operate, the last of Hong Kong's four plants to do so:[2]

- Lai Chi Kok Incineration Plant - commissioned 1969 and decommissioned 1991; now part of the expanded container port area

- Kennedy Town Incineration Plant - commissioned 1967 and located next to Island West Transfer Station; decommissioned 1993 and now berthing facility

- Mui Wo Incineration Plant - commissioned 1987 and decommissioned 1994; now site of Silvermine Bay Outdoor Recreation Camp

5.7.2   Demolition

Although the plant ceased operation in 1997, it was not completely demolished. The site was found to be contaminated with dioxin, furan, asbestos, heavy metals and petroleum hydrocarbons. Special procedures were required during demolition.

Demolition of the building and 150-metre-high (490 ft) chimney was initiated in 2007.[3]

The site cleanup preceded the demolition to clean up potential toxins and for future development of the site.[4]

5.7.3   See also

- Waste management in Hong Kong

5.7.4   References

[1] Incinerator The Best Option For Hong Kong Rubbish - Blog entry by Clear the Air, January 2008

[2] FAQ on incineration from the webpage of The Hong Kong Environmental Protection Department

[3] Kwai Chung Incineration Plant Demolition and Decontamination Works , Contract on the demolition

[4] http://www.cedd.gov.hk/eng/about/organisation/org_sdw.htm

Coordinates: 22°21′30″N   114°06′58″E / 22.3583°N 114.1160°E

5.8   Regional Council (Hong Kong)

The Regional Council's service area (in green)

The **Regional Council** (**RegCo**; Chinese: 區域市政局) was a municipal council in Hong Kong responsible for municipal services in the New Territories (excluding New Kowloon). Its services were provided by the Regional Services Department. Its headquarters were located near Sha Tin Station.

5.8.1   History

Technically, only Hong Kong Island, Kowloon, and New Kowloon were within the purview of the Urban Council. But the Urban Services Department, the executive arm of the Urban Council, began servicing the New Territories with its establishment in 1953.[1]

Following public consultation, a **Provisional Regional Council** was established on 1 April 1985 under the auspices of the colonial Hong Kong Government, to provide for the New Territories what the Urban Council did for Hong Kong Island and Kowloon.[1] Like the Urban Council, the Regional Council was created in 1986 as an elected body comprising representatives from constituencies and district boards.

In 1986, planning began for the council's headquarters building. Until permanent premises were built, departments of the Regional Council were scattered around various buildings in Tsim Sha Tsui.*[2] A site was selected near Sha Tin Town Centre and construction began in April 1989.*[3] It was opened on 27 September 1991 by governor David Wilson and Lady Wilson.*[2] The building consisted of a low block, housing the council chambers, alongside a 20-storey tower home to the various units of the Regional Services Department.*[4] The building was designed by Peter Keeping, a senior architect of the Architectural Services Department, and cost $200 million.*[3] The entrance is guarded by two marble lions made in Beijing. Today the building is the headquarters of the Leisure and Cultural Services Department.

5.8.2 Function and structure

The Regional Council structure comprised the full Regional Council, functional select committees, district committees, and sub-committees.

Initially, three functional select committees were planned: the Ways and Means Select Committee, the Environmental Hygiene Select Committee, and the Recreation and Culture Select Committee. They were joined by the Liquor Licensing Board at the founding of the council in 1986, and in 1987 the Ways and Means Select Committee was split into two committees: the Capital Works Select Committee and the Finance and Administration Select Committee.*[5] From 4 July 1997, the Recreation and Culture Select Committee was separated into the Culture and Arts Select Committee and the Recreation and Sports Committee, forming an eventual six select committees by the time the council was dissolved.*[6]

The nine district committees were as follows: Islands; Kwai Chung and Tsing Yi; North District; Sai Kung; Sha Tin; Tai Po; Tsuen Wan; Tuen Mun; and Yuen Long District Committee.

5.8.3 Demise

After the transfer of sovereignty in 1997, the name was once again changed to Provisional Regional Council, consisting of members of the pre-handover RegCo, and new members appointed by the Chief Executive. The council was dissolved on 31 December 1999 together with the Provisional Urban Council under the then-Chief Executive Tung Chee Hwa's plan to streamline and centralise municipal services as part of his government policy reforms. The Regional Council and Urban Council had, since 1998, jointly objected to this plan, putting forward an alternative merger proposal entitled "One Council, One Department", which

was not accepted by the government.*[6] The final chairman commented:

> "Subsequent to [the Council's] establishment, marked improvements had been made to the cultural, recreational, and entertainment services and facilities of the region, and they were highly regarded and cherished by the local community in the New Territories. It is therefore a great pity to see the dissolution of the council in such haste and by such a murky decision based on unconvincing arguments. Although it was clear to all of us that, with experiences acquired from serving the council for more than a decade, we could do more and better for the people of Hong Kong."
>
> —*Lau Wong-fat, Regional Council member from 1986–1999*[7]

The functions of the councils were replaced by two newly established government departments, the Food and Environmental Hygiene Department and the Leisure and Cultural Services Department. The former Regional Council Headquarters is now home to the Leisure and Cultural Services Department. The archives of the two municipal councils are held by the Hong Kong Public Libraries, and are available online in digitised form.*[8]

5.8.4 Chairmen

- Cheung Yan-lung (1986-1995)
- Lam Wai-keung (1995-1997)
- Lau Wong-fat (1997-1999)

5.8.5 Notes

[1] Lau 2002, p. 143.

[2] Chan, Catherine (26 September 1991). "Building for the future of Hongkong". *South China Morning Post*. p. 25.

[3] Chan, Catherine (26 September 1991). "New base for HK's essential services". *South China Morning Post*. p. 26.

[4] Liu, Mei-fong (August 1992). *A study of office decentralization in Sha Tin New Town*. University of Hong Kong. pp. 69–70.

[5] Lau 2002, pp. 143-44.

[6] Lau 2002, p. 150.

[7] Lau 2002, pp. xix.

[8] "Municipal Councils Archives Collection". Hong Kong Public Libraries. Retrieved 27 January 2015.

Bibliography

- Lau, Y.W. (2002). *A History of the Municipal Councils of Hong Kong 1883-1999*. Hong Kong: Leisure and Cultural Services Department. ISBN 962-7039-41-1.

5.8.6 External links

- Food and Environmental Hygiene Department

- Leisure and Cultural Services Department

5.9 Sai Tso Wan Recreation Ground

Sai Tso Wan Recreation Ground, evening view

Sai Tso Wan Recreation Ground (晒草灣遊樂場) is a multi-purpose playground in Lam Tin, Hong Kong. It is the first permanent recreational facility in Hong Kong built from a landfill.[1]

5.9.1 History

From 1978 to 1981, the knoll currently occupied by Sai Tso Wan Recreation Ground was known as Sai Tso Wan Landfill. The landfill served East Kowloon. During its operation, the landfill held approximately 1.6 million tonnes of domestic waste and commercial waste. Rubbish in the landfill stacked up to 65 metres high. After its closure in 1981, it was sealed with soil. The landfill then underwent a series of restoration works from 1995 to 2004, which turned the landfill into a recreation ground.

5.9.2 Construction

The ground was built from 1995 to 2004 from the former Sai Tso Wan Landfill. During the construction, the former landfill underwent a series of restoration works, which included the building of a final capping layer for prevention of leakage, a landfill gas control system for utilization of methane gas generated from the decomposed rubbish, and a leachate management system.

Sai Tso Wan Recreation Ground was commissioned on April 30, 2004. The ground has a multi-purpose sand-based, grass pitch which can be used for baseball and football activities, two (2) baseball batting cages, a children's playground, a jogging track, two changing rooms and a management office. Unlike most urban areas which relies mainly on electric supply, the ground is powered by wind turbines, solar cells and energy generated from combustion of methane gas.[2]

5.9.3 Location

Sin Fat Road near Sai Tso Wan Recreation Ground

Located adjacent to Lam Tin MTR Station, the ground is a common sporting destination of many residents in Lam Tin. Its carpark also allows residents outside the district to access the facility easily.[3]

5.9.4 Facilities

The recreation ground consists of a multi-purpose grass pitch for both football and baseball, two batting cages for baseball practising, a children's play area, a jogging track and a number ancillary facilities, including male and female bathrooms and an office. The ground is a training venue for the National Squad of the Hong Kong Baseball Association.

Its proximity to Lam Tin District made the ground a common sporting destination of Lam Tin residents.

Sai Tso Wan Park statue.

The ground opens from 7am to 11pm daily. Near its entrance is a main gate which has both a pedestrian ramp and a driveway, including an electric gate for the parking lot.*[4]

5.9.5   Environmental measures

Sai Tso Wan Recreation Ground was built in part to promote environmental protection in Hong Kong. Therefore, it was built with a number of features for environmental protection.

Wind turbines generate electricity by wind, which can either be used directly or stored in batteries for later use. Most electricity generated from the turbines is used for street lighting.

Solar panels in the recreation ground absorb sunlight during sunny days and convert the energy into electricity. The electricity generated from solar panels is first stored in batteries, and then used to power the fluorescent lights and electric fans in the reception area.

Surface water, including rainwater, collected from the facility is drained, recycled and then used to irrigate the turf of the ground.

Rubbersoil, a recycled, lightweight and porous material made from cement and shredded old tyres, is used as a sub-base material of the facility. Safety mats in the ground is also derived from used tyres.

A statue made from cement, crushed construction waste and glass pieces was erected in front of the ground's office as a landmark of the ground.

5.9.6   See also

- Waste management in Hong Kong

5.9.7   References

[1] "Sai Tso Wan Recreation Ground Open For Public Use" (Press release). Environmental Protection Department, Government of the Hong Kong Special Administrative Region. April 28, 2004. Retrieved 2007-03-08.

[2] *Sai Tso Wan Recreation Ground* - Environmental Protection Department. Retrieved on 13 February 2007.

[3] Centamap

[4] *Proposed Sai Tso Wan Recreation Ground on the Restored Sai Tso Wan Landfill* - Environmental Protection Department. Retrieved on 8 March 2007.

Coordinates: 22°18′17″N 114°13′57″E / 22.3046°N 114.2325°E

5.10   Sha Tin Sewage Treatment Works

Sha Tin Sewage Treatment Works (Chinese: 沙田污水處理廠) is a sewage treatment facility in Hong Kong. It is located in Ma Liu Shui, Sha Tin,*[1] along the Shing Mun River, at its mouth into Sha Tin Hoi (Tide Cove).

The treatment works serves Sha Tin, Ma On Shan and the villages nearby.*[2] The plant is managed by the Drainage Services Department.*[2]

It was then extended in several stages. Stage I was first commissioned in 1982*[2] with stage II following in 1986. Stage III was completed in September 2004.*[3]

Overview of Sha Tin Sewage Treatment Works. The Sha Tin Race-course is in the foreground. Sha Tin Hoi and Ma On Shan are in the background.

Sha Tin Sewage Treatment Works

5.10.1 References

[1] Drainage Services Department contact

[2] Sha Tin sewage treatment works open to the public

[3] Shatin Sewage Treatment Works Expansion and Upgrade

5.10.2 External links

- Drainage Services Department website

- Sha Tin Sewage Treatment Works, Stage III Environmental Impact Assessment Study

Coordinates: 22°24′24″N 114°12′49″E / 22.4066°N 114.2136°E

5.11   Urban Council

For other uses, see Urban Council (disambiguation).

The **Urban Council** (UrbCo) was a municipal council

Symbol of the Urban Council from its inception in the 1960s until its abolition in 1999.

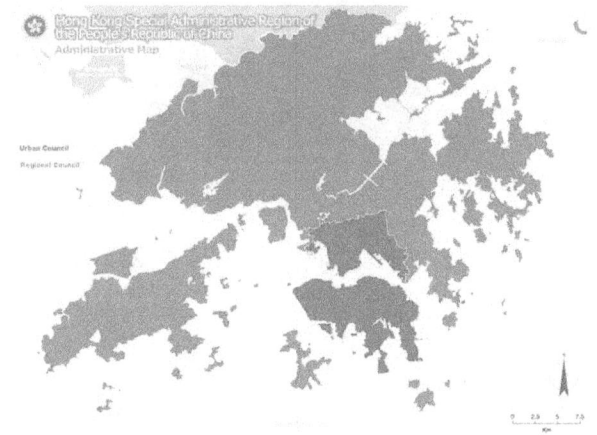

The Urban Council's service area (in pink).

in Hong Kong responsible for municipal services on Hong Kong Island and in Kowloon (including New Kowloon). These services were provided by the council's executive arm, the Urban Services Department. The equivalent body for the New Territories was the Regional Council.

The council was founded as the Sanitary Board in 1883. It was renamed the Urban Council when new legislation was passed in 1936 expanding its mandate. In 1973 the council was reorganised under non-government control and became financially autonomous. Originally comprised mainly of *ex officio* and appointed members, by the time the Urban Council was disbanded following the Handover it was comprised entirely of members elected by universal suffrage.

The Urban Council ran numerous public services including public libraries. Shown here is the logo of the Urban Council Public Library Reading Programme, a reading programme in the 1990s which provided gifts as incentives for children to read, based on the number of books they borrowed and read.

The Hong Kong Park was jointly built by the Urban Council and the Royal Hong Kong Jockey Club

5.11.1 History

The Urban Council was first established as the Sanitary Board in 1883. In 1887, a system of partial elections was established, allowing selected individuals to vote for members on the board. On 1 March 1935, the Sanitary Board was reconstituted to carry out the work which remained much the same until World War Two broke out. The board gained a new name in 1936 when the government passed the Urban Council Ordinance, legally expanding the range of services provided by the council, which had been gradually increasing in scope regardless.[1]

After the Second World War, the Urban Council received its pre-war form but without any elected members. The work of the Sanitary Department of the government began to separate out from the medical and health service. The first Urban Council meeting to take place after the Japanese occupation was held on 28 May 1946, with the council being empowered to carry out all its old duties – cleaning, burying the dead, running bath houses and public lavatories, hawker control – as well as some new ones, such as the use of bathing beaches throughout Hong Kong.

Only in May 1952 were elections returned to the Urban Council when two members were elected. And later in 1952, the number of elected members was doubled, their terms of office extended to two years and the electoral roll enlarged.

Finally by April 1956 half of the members of the Urban Council was elected by a small minority of the population eligible to vote. The qualifications for eligibility were very complex: For example, a voter had to be at least 21 years of age, have lived in Hong Kong for at least 3 years and must be qualified in at least one of 23 categories, which included educational qualifications (School Certificate Examination or equivalent), be a juror, salaried taxpayer, or a member of certain professional organisations. More details can be found in Schedule 1 of the *Urban Council Ordinance* (Cap. 101, *Laws of Hong Kong*).[2] It was estimated that in 1970 there were 250,000 eligible voters[3] and in 1981 the number had increased to 400,000 – 500,000.[4]

In 1960s, the responsibilities of the Urban Council continued to multiply. The City Hall in Central was opened in 1962, followed by the first multi-storey markets in Jardine's Bazaar in March 1963.

In 1973 the council was reorganised under non-government control with financial autonomy, which meant that the budget could be planned without the approval of the legislative council. Furthermore, the changes also removed housing as one of its main tasks. Since then, there were no government officials on the council and both the chairman and vice-chairman were elected among the 24 members. At that time, the council was the only one which solely consisted of members of the public.

Source: Norman Miners, 1986, *The Government and Politics of Hong Kong* p. 167.

Source: Norman Miners, *The government and politics of Hong Kong* (Hong Kong; New York: Oxford University Press, 1981), p. 224.

Prominent elected Urban Councilors were Elsie Tu and Brook Bernacchi of the Reform Club.

The Urban Council celebrated its centennial anniversary in 1983. The Urban Council Centenary Garden was named to commemorate the occasion.

An equivalent body, the Regional Council was set up in 1986 to serve the New Territories (excluding New Kowloon).

In 1994 the Council became fully elected based on universal and equal adult suffrage.[5]

After the transfer of sovereignty in 1997, the name was changed to **Provisional Urban Council**, consisting of members of the pre-handover UrbCo, and new members appointed by the Chief Executive.

5.11.2 Duties and services

Mong Kok Stadium

The former Urban Services Department Training School

The Urban Council provided a spectrum of services to the Hong Kong people over the years. The Urban Services Department was the executive branch of the council to imple-

Kowloon Walled City Park

ment policies and services. In 1997, it had about 16,000 employees, according to its published leaflet of 'service promises'.

The council's services included: recreational venues and activities, libraries, museums, cultural and entertainment venues, ticketing, wet markets, hawker registration and control, cremation, street cleansing, issuing licenses, and operating abattoirs.

Arts and culture

The Urban Council had played a significant role in the artistic and cultural development of Hong Kong.

It also managed the Urban Council Public Libraries system in Hong Kong Island and Kowloon which, upon the dissolution of the municipal councils, was merged with the Regional Council Public Libraries to form Hong Kong Public Libraries.

Cultural events   Since 1976, the council held its major cultural presentation – Festival of Asian Arts. The International Film Festival was another council-sponsored event, taking place annually mid-year and giving Hong Kong people a rare chance to see a range of international film-making, as well as Chinese films. The Independent Short Film and Video Awards were founded in 1993.

Museums   The Hong Kong Museum of Art gives regular exhibition of both Chinese and Western art and sculpture and frequently arranges art exchanges with overseas countries. The Hong Kong Museum of History, once temporarily housed in the Kowloon Park, featured the recording of local history and oral tradition. It is now located at Chatham Road in Tsim Sha Tsui. The Hong Kong Space Museum presents shows in the Space Theatre and exhibitions on astronomy, nature and space exploration with IMAX techniques.

Arts groups   The council directly financed and often even managed many local arts groups. In 1983, at "An Evening With the Council's Performing Companies" – one of the events in the Urban Council Centenary Celebration – the then-council chairman Hilton Cheong-Leen said, "Together with the Government, the Urban Council is committed to the development of the arts in Hong Kong. We aim to do so at the professional level so that gifted Hong Kong citizens can develop their artistic potential. We also aim to make available to all members of the community a wide range of artistic performance for their enjoyment and appreciation. And in the not too distant future we hope to see Hong Kong recognised as a major international centre of the performing arts."

The Hong Kong Chinese Orchestra was established in 1977, under direct financial support and management by the Urban Council.

The Hong Kong Repertory Theatre was also founded in 1977 and was directly financed and administered by the Urban Council, aiming to promote and raise the standards of the theatrical "stage play" drama in Cantonese in the territory with professional actors, directors, playwrights, administration, training and production.

The Hong Kong Dance Company was established in May 1981, and was at one time directly administered by the Urban Council. It aims to combine classical and folk traditions of China with contemporary international awareness. These groups were later taken over by the Leisure and Cultural Services Department when the Urban Council was dissolved. In 2001, the groups were privatised and became limited companies, but still receive funding from the government.

Recreation and sport

The council operated sports grounds, parks, indoor games halls, and public swimming pools.

Sanitation

See also: Hawkers in Hong Kong

The council was responsible for street cleansing, refuse collection, and pest control. It operated refuse collection points, public toilets and bathhouses, and was responsible for rubbish bins throughout the urban area. It was also responsible for the control of hawkers, issuing hawker licences and operating hawker bazaars.

5.11.3   Demise

In early 1997, chief executive-designate Tung Chee-hwa announced that the two municipal councils would be disbanded on 1 July 1997 (the Handover) and replaced by two provisional councils, with members appointed by the government, that would serve until elections in 1999. Tung said that those reappointed must "love China [and] love Hong Kong" and refused to clarify whether democratic politicians met this definition.[6][7][8] The announcement caused a row at the Urban Council and was unpopular with the public.[9] Likewise, the pre-handover government opposed China's decision to disband the two councils and the 18 district boards, and to reintroduce appointed seats, which had been abolished under democratic reforms.[10]

The post-handover provisional executive council met in May 1997 and drafted new legislation that would allow the chief executive-designate to appoint members to the new provisional bodies.[11] Three bills re-introducing appointed seats to the post-handover municipal councils and district boards were passed by the provisional legislature on 7 June 1997. The Urban Council, Regional Council and District Boards (Amendment) Bills 1997 stipulated that Tung Chee-hwa could appoint no more than 50 seats to the provisional municipal councils.[12] Frederick Fung, chairman of the ADPL, called the bills a "retrogression of democracy" while Chan Kam-lam of the pro-Beijing DAB asserted that "elections were divisive and appointments would stabilise the community".[12] Also on 1 July, elements of the Urban Council Ordinance and Regional Council Ordinance were repealed to allow the government to determine the composition and tenure of the councils.[13][14]

After the handover the council was disbanded and replaced with the Provisional Urban Council, which comprised pre-handover councillors plus new members appointed by the new government. The same was done with the Regional Council. The government then announced that the councils would be abolished in 1999. Both councils jointly objected to this plan, putting forward an alternative merger proposal entitled "One Council, One Department", which was not accepted by the government.[15]

Both councils were dissolved on 31 December 1999 as planned. Within days of the dissolution of the Urban Council, its distinctive symbol was systematically removed from public sight, such as by pasting over it with paper on all litter bins and information boards. Shortly afterwards, all the litter bins were themselves discarded, replaced by a similar design, but in green rather than purple. The duties of the councils were taken up by two newly-created government departments: the Food and Environmental Hygiene Department and the Leisure and Cultural Services Department.

The archives of the two municipal councils are held by the Hong Kong Public Libraries, and are available online in digitised form.*[16]

5.11.4 Chairmen

Before 1973, the chairmanship was occupied by the Director of Urban Services:

- A. de O. Sales, 1973–1981
- Hilton Cheong-Leen, 1981–1986
- H.M.G. Forsgate, 1986-1991
- Ronald Leung Ding-bong, 1991–1999

5.11.5 References

Citations

[1] Lau 2002, p. 32.

[2] Norman Miners. 1981. *The Government and Politics of Hong Kong*. Hong Kong: Oxford University Press.

[3] "Elected Urbco protest over reform plan," in: *South China Morning Post*, 1970

[4] "Sing Tao Jih Pao," in *Hong Kong Standard*, 8 March 1981

[5] CACV 1/2000

[6] No, Kwai-yan (13 March 1997). "No firm answer from Tung". *South China Morning Post*. p. 6.

[7] Li, Angela (17 March 1997). "Number of members for bodies yet to be decided". *South China Morning Post*. p. 5.

[8] "Tung adds condition for handover survivors". *South China Morning Post*. 2 February 1997. p. 2.

[9] Li, Angela (5 February 1997). "Let councillors stay, says poll". *South China Morning Post*. p. 6.

[10] Li, Angela (18 March 1997). "Legislators reveal concern at secondment". *South China Morning Post*.

[11] Hon, May Sin-mi; Li, Angela (7 May 1997). "Power to appoint in pipeline". *South China Morning Post*. p. 6.

[12] "Appointed seats bills passed". *South China Morning Post*. 8 June 1997. p. 4.

[13] "Laws to be scrapped". *South China Morning Post*. 20 January 1997. p. 4.

[14] "How the laws are affected". *South China Morning Post*. 21 January 1997. p. 6.

[15] Lau 2002, p. 150.

[16] "Municipal Councils Archives Collection". Hong Kong Public Libraries. Retrieved 27 January 2015.

Sources

- Lau, Y.W. (2002). *A History of the Municipal Councils of Hong Kong 1883–1999*. Hong Kong: Leisure and Cultural Services Department. ISBN 962-7039-41-1.

5.11.6 External links

- Food and Environmental Hygiene Department
- Leisure and Cultural Services Department

5.12 Waste management in Hong Kong

In the densely populated Hong Kong, waste is a troublesome issue. The territory generates around 6.4 million tons of waste each year,*[1] and by 2015, its existing landfills are expected to be full.*[2] The government has introduced waste management schemes and is working to educate the public on the subject. On the commercial side, producers are taking up measures to reduce waste.

5.12.1 Statistics

Hong Kong EPD (Environmental Protection Department) provides data and statistics about waste management.*[3]

5.12.2 Waste management process

Overview

In Hong Kong, wastes generated can be categorised as municipal solid waste, construction and demolition waste, chemical waste and other special waste, including: clinical waste, animal carcasses, livestock waste, radioactive waste, grease trap waste and waterworks/sewage sludges. Current (2012), according to Waste Atlas 1st Report waste generation in Hong Kong is around 3,3 million tonnes per year or 464 kg/cap/year.*[4]

Processing

Wastes in Hong Kong are first collected from disposal bins to refuse transfer stations (RTS). After they are compacted and put in containers, they are delivered to disposal lands or recycling centers.

Waste Collection

There are hundreds of collectors in the territory where wastes are located before transferring to refuse transfer stations.[5]

Waste Transport

There are seven refuse transfer stations in the territory. They serve as centralised collection points for the transfer of waste to the strategic landfills.[6]

Landfills

Operated by the EPD, the landfill sites only accept garbage from Hong Kong. Thirteen of 16 landfills were closed from 1988 to 1996.

Strategic landfills Hong Kong has three strategic landfills in use. All are located in the New Territories:[7][8][9]

Closed landfills There are also 13 closed landfills.[10] The closed landfills are converted into facilities such as golf courses, multi-purpose grass pitches, rest gardens, and ecological parks. Greenhouse gases emitted from closed landfills are used for energy. The closed landfills are:[8][10]

5.12.3 NGO campaigns

Friends of the Earth

Friends of the Earth (HK) is one of the local environmental groups in Hong Kong. One of its campaigns emphasises on setting up an all-inclusive recycling system.[11]

Green Power

Green Power, another local environmental organisation, has many activities related to waste control and management. Green Power organises an ongoing "Zero Waste Action", aiming to reduce the waste the territory produces.[12]

5.12.4 See also

- Air pollution in Hong Kong
- Environment of Hong Kong
- Sha Tin Sewage Treatment Works

5.12.5 References

[1] *Waste problem in Hong Kong* (PDF), retrieved 2009-06-28

[2] Ockenden, James (2007-02-27). "HK landfills full in 4-8 years". *blueskieschina.com*. Retrieved 2009-06-28.

[3] EPD - Data & Statistics

[4] "Waste Atlas. (2012). Country Data: HONG KONG SAR, CHINA." .

[5] Friends of the Earth (HK)

[6] http://sc.info.gov.hk/gb/www.epd.gov.hk/epd/english/environmentinhk/waste/prob_solutions/msw_rts.html

[7] Hong Kong Expands Municipal Solid Waste Management System Into The Future

[8] http://webcache.googleusercontent.com/search?q=cache:BzhHy_GrkQgJ:www.legco.gov.hk/yr05-06/english/sec/library/0506in37e.pdf+Landfills+in+Hong+Kong&hl=en&ct=clnk&cd=6&gl=ca

[9]

[10] http://sc.info.gov.hk/gb/www.epd.gov.hk/epd/english/environmentinhk/waste/prob_solutions/msw_racl.html

[11] http://www.foe.org.hk/welcome/geten.asp?language=en&id_path=1,%207,%2026,%203008,%203144—

[12] Green Power - Activities

5.12.6 External links

- Restoring Hong Kong's Landfills
- Hong Kong Environmental Protection Department
- Friends of the Earth
- Green Power

Chapter 6

Related Concepts & Terms

6.1 Animal welfare and rights in China

Animal welfare and rights in China is a topic of growing interest and the ideas of animal welfare and animal rights were introduced to China in the 1990s. Animal-rights activists frequently condemn China's treatment of animals. Movements towards animal welfare and animal rights are expanding in China, including among homegrown Chinese activists.

6.1.1 Legislation

China currently has no animal-welfare laws.[*][1][*][2][*][3]

In 2006, Zhou Ping of the National People's Congress introduced the first nationwide animal-protection law in China, but it didn't move forward.[*][4]

In September 2009, the first comprehensive Animal protection law of the People's Republic of China was introduced, but it hasn't made any progress.[*][3]

6.1.2 History

Several traditional Chinese worldviews emphasize caring for animals, including Taoism and Buddhist vegetarianism.[*][1] Taoist Zhuang Zhou taught compassion for all sentient beings.[*][5]

In more recent times, Prof. Peter J. Li suggests, many in mainland China have become relatively indifferent to animal suffering, perhaps partly because of Mao Zedong's campaigns against bourgeois sentiments, such as "sympathy for the downtrodden".[*][1] Caring about animals was regarded as "counter-revolutionary".[*][6] Since 1978, China has emphasized growth and avoidance of famine, which the government considers important for political stability. Local officials are evaluated based on local jobs and revenue. This has led to less concern for animal welfare.[*][1]

6.1.3 Consumption of animals

Livestock

Livestock farming has grown exponentially in China in recent years, such that China is now "the world's biggest animal farming nation."[*][7] In 1978, China collectively consumed 1/3 as much meat as the United States. By 1992 China had caught up, and by 2012, China's meat consumption was more than double that of the U.S.[*][8]

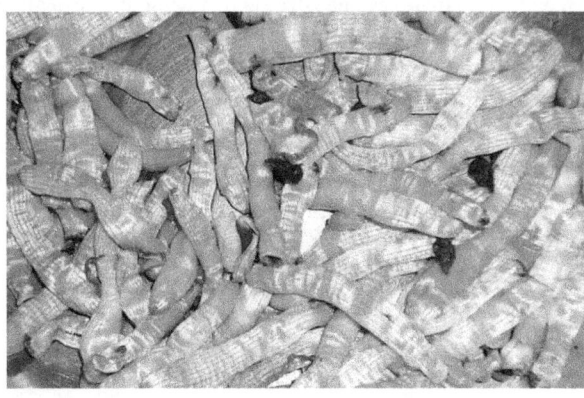

Marine animals in a market in Hainan Province, China

Almost 3/4 of China's meat is pork, and China's 476 million pigs comprise half of the world's pig population.[*][8] China produces 37 million tons of farmed fish—more than 60% of the world's total.[*][8]

A 2005-2006 survey by Prof. Peter J. Li found that many farming methods that the European Union is trying to reduce or eliminate are commonplace in China, including gestation crates, battery cages, foie gras, early weaning of cows, and clipping of ears/beaks/tails.[*][1] Livestock in China may be transported over long distances, and there are currently no humane-slaughter requirements.[*][1]

Cooking animals live

In 2008, more than 40 animal activists in Beijing gathered to protest skinning and cooking live cats in Guangdong province.[9] A 2010 article featuring content from Tiexue and Mop news sources showed pictures of skinned cats being submerged in boiling water.[10]

"The worse you treat them the better they taste. It makes sure the blood gets into the meat and it tastes delicious."

A cook referring to cats in the documentary *San Hua*[11]

The 2010 documentary *San Hua* by Guo Ke is the first to depict China's cat-meat industry. In one scene, Guo and fellow activists stop a transport truck and find "more than 300 cats crammed into cramped wooden cages, unable to move"—some missing tails and others "crushed into unconsciousness." In another scene at Fa's Cat Restaurant, Guo used a hidden camera to film cooks beating cats with a wooden stick, dumping them into a fur-removal machine, and then boiling them.[11]

Pictures have also circulated featuring two dogs in boiling water in China. It is claimed that this is because some Chinese prefer the taste of adrenaline-soaked meat. In some areas, dogs are beaten to death in order to release blood into the meat.[12]

Yin Yang fish involves deep-frying fish while it is still alive. The practice has been condemned by animal-rights activists. Many chefs in Taiwan are no longer willing to prepare it, but it is popular in mainland China.[13]

Some chefs cook a carp's body while keeping its head wrapped in a cloth so that it can keep breathing. In 2009, a video of Chinese diners prodding and eating alive a fried fish went viral on YouTube and provoked an outcry from PETA.[14]

On streets in China, live scorpions are "scooped up alive and wriggling, skewered on a kebab, and deep-fried in oil." [15]

Eating animals live

Main article: Eating live seafood

Drunken shrimp are eaten while struggling to get away.[16] One tourist visiting China described eating drunken shrimp as follows: "Everyone at the table reached into the bowl, chose a particularly feisty little (or rather quite big) shrimp, and placed him on their plate. As poor Mr. Shrimp jumped up and down [...] you picked him up, ripped off his head, and proceeded to peel him as fast as you can." [17]

Some Chinese food markets include live animals, such as live scorpions.[18]

6.1.4 Non-meat farming

Bile bears

Main article: Bile bear

China farms about 10,000 Asiatic black bears for bile

A black bear inside a "crush cage" made for bile extraction.

production—an industry worth roughly $1.6 billion per year.[1] The bears are permanently kept in cages, and bile is extracted from cuts in their stomachs.[1] In Jan. 2013, Animals Asia Foundation rescued six bile bears, which had broken and rotted teeth due to gnawing at their cages.[19]

Jackie Chan and Yao Ming have publicly opposed bear farming.[19][20][21] In 2012, over 70 Chinese celebrities took part in a petition against an IPO application by Fujian Guizhentang Pharmaceutical Co. due to the company's selling of bear-bile medicines.[22] In 2013, the company pulled its IPO application.[23]

According to Jill Robinson, over 1000 Chinese medicine stores have committed to not selling bear bile, but this compares with over 40,000 such shops in all of China.[24]

Fur

China is the biggest fur-producing nation.[1] Some fur animals are skinned alive, and others may be beaten to death with sticks.[1]

In Nov. 2013, PETA released a video of a live angora rabbit in northeastern China having its fur torn off. The video received 200,000 views on China's video site Youku within a month and prompted UK retailers like Primark and Topshop to stop imports from China of products using angora wool.[25]

Other

A caged Civet cat.

Asian palm civets are farmed in battery cages to produce Kopi Luwak ("civet coffee").

6.1.5 Animal testing

China has a $32 billion beauty market, and over 300,000 animals are thought to be used each year for required product tests.*[26] China is the only major buyer where mascaras and lotions need to be tested on animals.*[26]

"Our R&D isn't as sophisticated, and the consumer here doesn't think as much about ideals such as animal testing. They care about the price, the brand, and the product."

Xu Jingquan, secretary general at the All-China Federation of Industry and Commerce, Beauty Culture & Cosmetics Chamber*[26]

In 2013, the China Food and Drug Administration (CFDA) relaxed its testing requirements by allowing Chinese companies to verify safety using data from overseas tests, including non-animal tests. Foreign companies are still required to perform animal testing, but Humane Society International was hopeful about further humane reforms to come.*[24]

On 30 June 2014, CFDA eliminated its requirement for animal testing of "ordinary cosmetics" like shampoos and some skin-care items as long as companies provided alternative data showing safety. This change does not extend to imported cosmetics or to "any special-use products, including hair dyes and sunblocks." *[27]

Some animal tests are likely to continue for now even on exempt products because some testers do not have the technology for alternative in vitro methods. Animal activists were excited by the announcement, and over 50 of them took

to the streets of Dalian in northeastern China to celebrate, wearing bunny ears.*[27]

6.1.6 Zoos

According to Prof. Peter J. Li, a few Chinese zoos are improving their welfare practices, but many remain "outdated", have poor conditions, use live feeding, and employ animals for performances.*[1] Safari parks may feed live sheep and poultry to lions as a spectacle for crowds.*[4]

6.1.7 Other animal-rights issues

Live-animal key rings

In Beijing, vendors sell fish, turtles, and amphibians as key rings and mobile-phone decorations. Animal-rights activists condemn the practice because the animals may run out of air and die quickly, and they may also pose hazards to human health.*[28]*[29]

6.1.8 Animal-rights movement in China

Ideas of animal welfare and animal rights were introduced to China in the 1990s.*[30]

"In many ways, the animal welfare movement in China is maturing far faster than it ever did in the West."

Jill Robinson*[2]

China's animal-protection movement is growing,*[30] particularly among young people,*[31] especially those in urban areas and on the Internet.*[6] International NGOs played some role in igniting China's animal movement, but local groups are increasingly taking over.*[21]

China is home to 130 million dogs, mostly pets.*[1] As China becomes wealthier, more people are owning pets, which increases opposition to animal cruelty.*[12] In Apr. 2012, activists rescued 505 dogs that were headed to slaughter from a truck where they had endured harsh conditions.*[12]

Chinese activists prevented introduction of a bullfighting project in 2010 and rodeos in 2011.*[1] Activists have preempted a foie gras factory, ended live feeding in zoos, and rescued thousands of dogs and cats from being killed for meat.*[2] Vegetarian restaurants are increasing, though partly because of fashion rather than ethics.*[4]

A 2011 survey of about 6000 Chinese found that while about 2/3 of respondents had never previously heard of "animal welfare", 65.8% expressed at least partial support of animal-welfare laws, and more than half said they were fully or partially willing to pay more for humane animal products.*[32]

6.1.9 Criticism of animal welfare in China

Tsinghua University professor Zhao Nanyuan argues that animal rights represents a form of Western imperialism ("foreign trash") that is "anti-humanity". He argues that animals are not sentient and therefore don't have rights. He encourages China to learn from the example of South Koreans who refused Western protests of its dog-meat traditions.*[33]

Critics have pointed out that while non-human animals are not as advanced in their needs and desires as humans, they do share some basic needs, such as food, water, shelter and companionship.

Some claim that it is contradictory for the U.S. to condemn China's mistreatment of animals while engaging in its own forms of animal cruelty. Chinese animal-welfare groups censured an American-style rodeo, as well as Jackie Chan's support for it. One Chinese commenter said of Chan: "You made a video about the protection of bears, and now you're promoting the mistreatment of cattle, it's a massive contradiction. Brother Chan, you've hurt me deeply."*[34]

6.1.10 See also

- Chinese Animal Protection Network
- Lychee and Dog Meat Festival, held each June
- Dog meat in China
- Wang Yan (activist), dog rescuer

6.1.11 Notes

[1] Tobias, Michael Charles (2 Nov 2012). "Animal Rights In China". Forbes. Retrieved 15 July 2014.

[2] Robinson, Jill (7 Apr 2014). "China's Rapidly Growing Animal Welfare Movement". Huffington Post. Retrieved 15 July 2014.

[3] Tatlow, Didi Kirsten (6 Mar 2013). "Amid Suffering, Animal Welfare Legislation Still Far Off in China". New York Times Blogs. Retrieved 15 July 2014.

[4] "A small voice calling". The Economist. 28 Feb 2008. Retrieved 15 July 2014.

[5] Miyun Park; Peter Singer (Mar–Apr 2012). "The Globalization of Animal Welfare". Foreign Affairs. Retrieved 15 July 2014.

[6] Levitt, Tom (26 Feb 2013). "Younger generation face long wait for law-change on animal cruelty". chinadialogue. Retrieved 20 July 2014.

[7] Li, Peter J. (Jun 2009). "Exponential Growth, Animal Welfare, Environmental and Food Safety Impact: The Case of China's Livestock Production". *Journal of Agricultural and Environmental Ethics* **22** (3): 217–240. doi:10.1007/s10806-008-9140-7. ISSN 1573-322X.

[8] Larsen, Janet (24 Apr 2012). "Meat Consumption in China Now Double That in the United States". *Earth Policy Institute*. Retrieved 15 July 2014.

[9] "China protesters: Stop 'cooking cats alive'". Associated Press. 18 Dec 2008. Retrieved 16 July 2014.

[10] Fauna (12 Oct 2010). ""Boiled Alive Cat" Prepared, Served In Guangzhou Restaurants". *chinaSMACK*. Retrieved 16 July 2014.

[11] Yiyan, Zhou (6 Oct 2010). "The fight to protect China's cats". chinadialogue. Retrieved 16 July 2014.

[12] Cooper, Rob (25 Jun 2012). "Dogs destined for the table: Horrific images show animals being killed, cooked and served up as a meal in Chinese tradition". Daily Mail. Retrieved 16 July 2014.

[13] "Chefs refuse to serve 'dead-and-alive fish' ". The China Post. 9 Jul 2007. Retrieved 16 July 2014.

[14] Leach, Ben (18 Nov 2009). "Chinese diners eat live fish in YouTube video". The Telegraph. Retrieved 16 July 2014.

[15] Greenfield, Beth (21 Oct 2011). "15 Insects You Won't Believe Are Edible". Budget Travel. Retrieved 1 June 2014.

[16] Griffin, Simon (6 Mar 2013). "10 Animals That People Eat Alive". *Listverse*. Retrieved 16 July 2014.

[17] Pastey White Guy (22 Jul 2006). "Drunken shrimp". *LBS: A Pastey White Guy's Perspective*. Retrieved 16 July 2014.

[18] Jou, Eric (17 Mar 2014). "It's The Snacktaku Fried Insect Special". *Kotaku*. Retrieved 16 July 2014.

[19] Yee, Amy (28 Jan 2013). "Market for Bear Bile Threatens Asian Population". New York Times. Retrieved 15 July 2014.

[20] "Jackie Chan PSA on Bear Bile Farming". World Animal Protection US. Retrieved 15 July 2014.

[21] "Animal Rights In China Get Boost From Celebrity Activists And Shifting Attitudes". Huffington Post. 22 Apr 2012. Retrieved 15 July 2014.

[22] Loo, Daryl (16 Feb 2012). "Chinese Celebrities Oppose IPO for Operator of Bear-Bile Farm". Bloomberg Businessweek. Retrieved 16 July 2014.

[23] Turk, Gregory (4 Jun 2013). "China Bear-Bile Farm Operator Among 269 Companies to Pull IPO". Bloomberg Businessweek. Retrieved 16 July 2014.

[24] Einhorn, Bruce (14 Nov 2013). "Animal-Rights Activists Celebrate Small Victories in China". Bloomberg Businessweek. Retrieved 16 July 2014.

[25] Gao, Helen (23 Jan 2014). "Letter from Beijing: Animal cruelty is rife in China—but things are changing". Prospect Magazine. Retrieved 15 July 2014.

[26] Lin, Liza (26 Sep 2013). "An Ugly Dilemma for Beauty Companies". Bloomberg Businessweek. Retrieved 16 July 2014.

[27] Huang, Shaojie (30 Jun 2014). "China Ends Animal Testing Rule for Some Cosmetics". New York Times Sinosphere. Retrieved 17 July 2014.

[28] "Live animals sold as key rings in China". CNN. 15 Apr 2011. Retrieved 16 July 2014.

[29] "Shell Shock". Snopes. 29 May 2014. Retrieved 16 July 2014.

[30] Lu, Jiaqi; Bayne, Kathryn; Wang, Jianfei (Nov 2013). "Current status of animal welfare and animal rights in China". Altern Lab Anim 41 (5): 351–357. PMID 24329743.

[31] McGuinness, Michelle (22 May 2013). "Animal welfare activists in China rise up against cruelty". MSN News. Retrieved 15 July 2014.

[32] Xiaolin You; Yibo Li; Min Zhang; Huoqi Yan; Ruqian Zhao (14 Oct 2014). "A Survey of Chinese Citizens' Perceptions on Farm Animal Welfare". PLoS ONE: e109177. doi:10.1371/journal.pone.0109177.

[33] Li, Peter J. (2006). "The Evolving Animal Rights and Welfare Debate in China: Political and Social Impact Analysis". Animals, Ethics and Trade: The Challenge of Animal Sentience. London: Earthscan. pp. 111–128.

[34] "Chinese claim Americans cruel to animals (Jackie Chan doesn't)". May Daily. 25 Jul 2011. Retrieved 20 July 2014.

6.2 Environmental policy in China

Environmental policy in China is set by the National People's Congress and managed by the Ministry of Environmental Protection. The Center for American Progress has described China's environmental policy as similar to that of the United States before 1970. That is, the central government issues fairly strict regulations, but the actual monitoring and enforcement is largely undertaken by local governments that have greater interest in economic growth. The environmental work of non-governmental forces, such as lawyers, journalists, and non-governmental organizations, is limited by government regulations.[1]

China's rapid economic expansion combined with the country's relaxed environmental oversight has caused a number of ecological problems. In response to public pressure, the national government has undertaken a number of measures to curb pollution in China and improve the country's environmental situation.[2] However, the government's response has been criticized as inadequate.[3] Encouraged by national policy that judges regions solely by their economic development, corrupt and unwilling local authorities have hampered enforcement.[4][5] Nonetheless, in April 2014, the government amended its environmental law to better fight pollution.[5]

6.2.1 Policy jurisdiction

The Ministry of Environmental Protection (MEP), formerly the State Environmental Protection Administration (SEPA), is a cabinet-level ministry in the executive branch of the Chinese Government that is responsible for implementing environmental policies, as well as the enforcement of environmental laws and regulations.[6] The Ministry is tasked with protecting China's air, water, and land from pollution and contamination. Directly under the State Council, it is empowered and required by law to implement environmental policies and enforce environmental laws and regulations. Complementing its regulatory role, it funds and organizes research and development.[7] There are 12 offices and departments under MEP, all at the si (司) level in the government ranking system. They carry out regulatory tasks in different areas and make sure that the agency is functioning accordingly. Since 2006, there have been five regional centers to help with local inspections and enforcement.

6.2.2 History

In 1972, Chinese representatives attended the first United Nations Conference on the Human Environment. The next year, the Environmental Protection Leadership Group was established. In 1983, the Chinese government announced that environmental protection would become a state policy. In 1998, China went through a disastrous year of serious flooding, and the Chinese government upgraded the Leading Group to a ministry-level agency, which then became the State Environmental Protection Administration.

According to the Chinese government website, the Central Government invested more than 40 billion yuan between 1998 and 2001 on protection of vegetation, farm subsidies, and conversion of farm to forests.[8] Between 1999 and 2002, China converted 7.7 million hectares of farmland into forest.[9]

From 2001 to 2005, Chinese environmental authorities received more than 2.53 million letters and 430,000 visits by 597,000 petitioners seeking environmental redress.[10]

Meanwhile, the number of mass protests caused by concerns over environmental issues grew steadily from 2001 to 2007.[11][12] The increased attention on environmental matters caused the Chinese government to display an increased level of concern towards environmental issues. For example, in his 2007 annual address Wen Jiabao, the Premier of the People's Republic of China, made 48 references to "environment," "pollution," and "environmental protection", and stricter environmental regulations were subsequently implemented. Subsidies for some polluting industries were cancelled, while other polluting industries were shut down. However, many internal environmental targets were missed.[4]

After the 2007 address, the influence of corruption was a hindrance to effective enforcement, as local authorities ignored orders and hampered the effectiveness of central decisions. In response, CPC General Secretary Hu Jintao implemented the "Green G.D.P." project, where China's gross domestic product was adjusted to compensate for negative environmental effects; however, the program quickly lost official influence due to unfavorable data. The project's lead researcher claimed that provincial leaders "do not like to be lined up and told how they are not meeting the leadership's goals ... They found it difficult to accept this." [4] The government attempted to hold national "No Car Days" where cars were banned from central roads, but the action was largely ignored.[13] In 2008, the State Environmental Protection Administration was official replaced by the Ministry of Environmental Protection during the March National People's Congress sessions in Beijing.[14]

Citizen activism regarding government decisions that are perceived as environmentally damaging increased in the 2010s.[15] In April 2012, protests occurred in the southern town of Yinggehai following the announcement of a power plant project. The protesters initially succeeded in halting the project, worth 3.9 billion renminbi (£387m). Another town was selected for the location of the plant, but when the residents in the second location also resisted the authorities returned to Yinggehai. A second round of protests occurred in October 2012 and police clashed with protester, leading to 50 arrests and almost 100 injuries.[16] In response to a waste pipeline for a paper factory in the city of Qidong, several thousand demonstrators protested in July 2012. Sixteen of the protesters were sentenced to between twelve and eighteen months in prison; however, thirteen were granted a reprieve on the grounds that they had confessed and repented.[17] In total, more than 50,000 environmental protests occurred in China during 2012.[18]

In response to an increasing air pollution problem, the Chinese government announced a five-year, US$277 billion plan to address the issue in 2013. Northern China will receive particular attention, as the government aims to reduce air emissions by 25 percent by 2017, compared with 2012 levels.[19]

In March 2014, CPC General Secretary Xi Jinping "declared war" on pollution during the opening of the National People's Congress.[20] After extensive debate lasting nearly two years, the parliament approved a new environmental law in April. The new law empowers environmental enforcement agencies with great punitive power, defines areas which require extra protection, and gives independent environmental groups more ability to operate in the country.[5] The new articles of the law specifically address air pollution, and call for additional government oversight.[20] Lawmaker Xin Chunying called the law "a heavy blow [in the fight against] our country's harsh environmental realities, and an important systemic construct" .[5] Three previous versions of the bill were voted down. The bill is the first revision to the environmental protection law since 1989.[5]

6.2.3 Current law

When the new environmental protection provisions go into effect in January 2015, the government's environmental agencies will be allowed to enforce strict penalties and seize property of illegal polluters. Companies that break the law will be "named and shamed" , with company executives subject to prison sentences of 15 days. There will be no upper limit on fines; previously, it was often cheaper for companies to pay the meager fines provisioned by the law than install anti-pollution measures. In all, the new law has 70 provisions, compared to the 47 of the existing law.[20] More than 300 different groups will be able to sue on the behalf of people harmed by pollution.[5] It remains to be seen whether these changes to the law will overcome some of the traditional problems with environmental litigation in China, such as difficulty getting cases accepted by the court, trouble gathering evidence and interference from local government.[21]

Under the new law, local governments will be subject to discipline for failing to enforce environmental laws. Regions will no longer be judged solely on their economic progress, but instead must balance progress with environmental protection. Additionally, local governments will be required to disclose environmental information to the public. Individuals are encouraged to "adopt a low-carbon and frugal lifestyle and perform environmental protection duties" such as recycling their garbage under the law.[20]

Protected areas

See also: List of protected areas of China

A number of different classes of protected areas are recognized under Chinese law. National, provincial, and local governments all have the power to designate areas as protected. Regardless of designation, most enforcement is made at the local level.

6.2.4   Current issues

Main article: Environmental issues in China

China has many environmental issues, severely affect-

Air pollution caused by industrial plants

ing its biophysical environment as well as human health. The water resources of China are affected by both severe water quantity shortages and severe water quality pollution. An increasing population and rapid economic growth as well as lax environmental oversight have increased water demand and pollution. China has responded by measures such as rapidly building out the water infrastructure and increased regulation as well as exploring a number of further technological solutions. Water usage by its coal-fired power stations is drying-up Northern China.[22][23][24] Desertification remains a serious problem, consuming an area greater than that taken by farmlands. Although desertification has been curbed in some areas, it still is expanding at a rate of more than 67 km^2 every year. 90% of China's desertification occurs in the west of the country.[25]

Various forms of pollution have increased as China has industrialized, causing widespread environmental and health problems.[26] In January 2013, fine airborne particulates rose as high as 993 micrograms per cubic meter in Beijing, compared with World Health Organization guidelines of no more than 25.[27] Heavy industry, dominated by state-owned enterprises, has been promoted since the beginning of central planning and still has many special privileges such as access to cheap energy and loans.[28] The industry possesses considerable power to resist environmental regula-

tion.[29]

6.2.5   Impact

China's lax environmental oversight has contributed to its environmental problems. Sixteen of the world's twenty most polluted cities are found in China.[2][27] Government response has been criticized as inadequate.[3] An official report released in 2014, found that 20% of the country's farmland, and 16% of its soil overall, is polluted. An estimated 60% of the groundwater is polluted.[20]

According to the U.S. Environmental Protection Agency, China has shown great determination to "develop, implement, and enforce a solid environmental law framework". However, the impact of such efforts is not yet clear.[30] The harmonization of Chinese society and the natural environment is billed as one of the country's top national priorities.[31]

International groups called the law revision passed in April 2014 a positive development, but cautioned seeing the laws through to implementation would be a challenge.[5]

Because China does not have a fully established legal system, enterprise executives base their environmental practices largely on perceptions about regulators rather than concerns for legal issues, according to a 2014 study published in Journal of Public Administration Research and Theory. One executive interviewed said that China's environmental regulations were "comprehensive" but yet "vague," leaving local officials with large discretion in terms of enforcement. If executives think local officials may arbitrarily target their enterprises for enforcement, they are likely to adopt proactive practices, such as "developing certifiable environmental management systems," but not basic ones, such as waste recycling.[32]

6.2.6   See also

- Climate change in China

- Debate over China's economic responsibilities for climate change mitigation

- Environmental governance in China

6.2.7   References

[1] Melanie Hart; Jeffrey Cavanagh (20 April 2012). "Environmental Standards Give the United States an Edge Over China". *Center for American Progress.* Center for American Progress. Retrieved 28 July 2013.

[2] "Air Pollution Grows in Tandem with China's Economy". NPR. Retrieved 2013-09-08.

[3] *China Weighs Environmental Costs; Beijing Tries to Emphasize Cleaner Industry Over Unbridled Growth After Signs Mount of Damage Done* July 23, 2013

[4] Joseph Kahn; Jim Yardley (26 August 2007). "As China Roars, Pollution Reaches Deadly Extremes". *The New York Times.* Retrieved 28 July 2013.

[5] "China Revises Environmental Law". *Voice of America.* 25 April 2014. Retrieved 27 April 2014.

[6] "华建敏：组建环境保护部加大环境保护力度 _ 新闻中心 _ 新浪网" (in Chinese). News.sina.com.cn. 11 March 2008. Retrieved 12 February 2013.

[7] Archived December 20, 2007, at the Wayback Machine.

[8] "Protection of forests and control of desertification". Retrieved April 2, 2008.

[9] Li, Zhiyong. " A policy review on watershed protection and poverty alleviation by the Grain for Green Programme in China". Retrieved April 2, 2008.

[10] Alex Wang. "Environmental protection in China: the role of law". China Dialogue. Retrieved 2013-02-12.

[11] Ma Jun (31 January 2007). "How participation can help China's ailing environment". *ChinaDialogue.* ChinaDialogue. Retrieved 28 July 2013.

[12] "Environmental Activists Detained in Hangzhou". *Human Rights in China.* Human Rights in China. 25 October 2012. Retrieved 28 July 2013.

[13] "China is holding a No Car Day in more than 100 cities as it tries to reduce smog ahead of the 2008 Summer Olympics." . *BBC News.* 22 September 2007. Retrieved 28 July 2013.

[14] "华建敏：组建环境保护部加大环境保护力度 _ 新闻中心 _ 新浪网". News.sina.com.cn. Retrieved 2013-02-12.

[15] Keith Bradsher (July 4, 2012). "Bolder Protests Against Pollution Win Project's Defeat in China". *The New York Times.* Retrieved July 5, 2012.

[16] "Chinese protesters clash with police over power plant". *The Guardian.* 22 October 2012. Retrieved 28 July 2013.

[17] "China jails anti-pollution protesters after riot". *The Age.* Reuters. 7 February 2013. Retrieved 28 July 2013.

[18] John Upton (8 March 2013). "Pollution spurs more Chinese protests than any other issue". *Grist.org.* Grist Magazine, Inc. Retrieved 28 July 2013.

[19] John Upton (25 July 2013). "China to spend big to clean up its air". *Grist.org.* Grist Magazine, Inc. Retrieved 27 July 2013.

[20] Jennifer Duggan (25 April 2014). "China's polluters to face large fines under law change". *The Guardian.* Retrieved 27 April 2014.

[21] Rachel E. Stern, Environmental Litigation in China: A Study in Political Ambivalence (Cambridge University Press, 2013); Xia Jun, "C China's Courts Fail the Environment," China Dialogue, January 16, 2012

[22] *Water Demands of Coal-Fired Power Drying Up Northern China* March 25, 2013 Scientific American

[23] *On China's Electricity Grid, East Needs West—for Coal* March 21, 2013 BusinessWeek

[24] *Chinese Utilities Face $20 Billion Costs Due to Water, BNEF Says* March 24, 2013 BusinessWeek

[25] HAN, Jun "EFFECTS OF INTEGRATED ECOSYSTEM MANAGEMENT ON LAND DEGRADATION CONTROL AND POVERTY REDUCTION." *Workshop on Environment, Resources and Agricultural Policies in China,* June 19, 2006. Retrieved March 26, 2008.

[26] Edward Wong (March 29, 2013). "Cost of Environmental Damage in China Growing Rapidly Amid Industrialization" . *The New York Times.* Retrieved March 30, 2013.

[27] Bloomberg News (14 January 2013). "Beijing Orders Official Cars Off Roads to Curb Pollution". *Bloomberg.* Retrieved 27 July 2013.

[28] "A new epic". *The Economist* (The Economist Newspaper Ltd). 21 October 2010. Retrieved 28 July 2013.

[29] Edward Wong (March 21, 2013). "As Pollution Worsens in China, Solutions Succumb to Infighting". *The New York Times.* Retrieved March 22, 2013.

[30] EPA, China Environmental Law Initiative.

[31] NRDC, Environmental Law in China.

[32] "Compliance with environmental regulations when the rule of law is weak: Evidence from China". JournalistsResource.org, retrieved July 30, 2014

6.3 One-child policy

The **one-child policy**, a part of the **family planning policy**, was a population control policy of China. It was introduced between 1978 and 1980 and began to be formally phased out in 2015. The policy allowed many exceptions and ethnic minorities were exempt. In 2007, 36% of China's population was subject to a strict one-child restriction, with an additional 53% being allowed to have a second child if the first child was a girl. Provincial governments imposed fines for violations, and the local and national governments created commissions to raise awareness and carry out registration and inspection work.

Demographers are not clear how much reduction happened solely because of the policy. According to the Chinese government, 400 million births were prevented, but other sources may vary between no effect and some hundreds of millions of prevented births. Although 76% of Chinese supported the policy in 2008,[1] it is controversial outside of China.

On 29 October 2015, it was reported that the existing law would be changed to a two-child policy, citing a statement from the Communist Party of China. The new law became effective from 1 January 2016, following its passage in the standing committee of the National People's Congress on 27 December 2015.

6.3.1 History

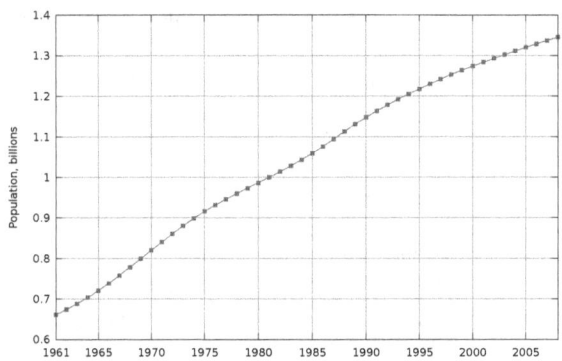

China's population (1961–2008)

Introduction

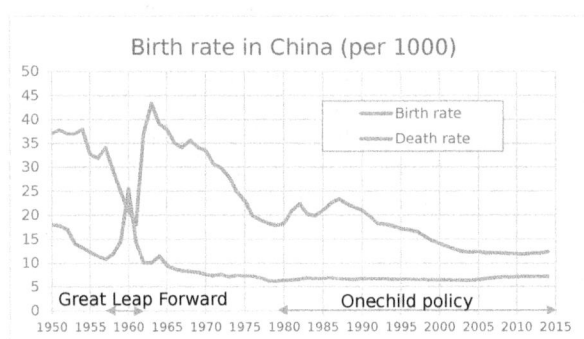

Birth rate in China

During the period of Mao Zedong's leadership in China, the birth rate fell from 37 per thousand to 20 per thousand,[2] infant mortality declined from 227 per thousand births in 1949 to 53 per thousand in 1981, and life expectancy dramatically increased from around 35 years in 1948 to 66 years in 1976.[2][3] Until the 1960s, the government encouraged families to have as many children as possible[4]

because of Mao's belief that population growth empowered the country, preventing the emergence of family planning programs earlier in China's development.[5] The population grew from around 540 million in 1949 to 940 million in 1976.[6] Beginning in 1970, citizens were encouraged to marry at later ages and have only two children.

Although the fertility rate began to decline, the Chinese government observed the global debate over a possible overpopulation catastrophe suggested by organisations such as Club of Rome and Sierra Club. While visiting Europe in 1979, one of the top Chinese officials, Song Jian, got in touch with influential books of the movement, The Limits to Growth and A Blueprint for Survival. With a group of mathematicians, Song determined the correct population of China to be 700 million. A plan was prepared to reduce China's population to the desired level by 2080, with the one-child policy as one of the main instruments of social engineering.[7] In spite of some criticism inside the party, the plan (also referred to as the Family Planning Policy[8]) was officially adopted in 1979.[9][10][11] The plan called for families to have one child each in order to curb a then-surging population and limit the demands for water and other resources,[12] as well as to alleviate social, economic and environmental problems in China.[13] The policy was formally implemented as a temporary measure on September 18, 1980.[14]

Relaxation

The one-child policy was originally designed to be a One-Generation Policy.[15] It was enforced at the provincial level and enforcement varied; some provinces had relaxed the restrictions. After Henan loosened the requirement, the majority of provinces and cities[16][17] permitted two parents, who were themselves the sole children of their respective parents, to have two children. As a result, in 2007, only 36% of China's population fell under the enforcement this policy.[18] In rural areas, however, families are allowed two children without incurring penalties.[19] The one-child limit has mostly been enforced in densely populated urban areas, and implementation varies from location to location.[20]

Beginning in 1987, official policy granted local officials the flexibility to make exceptions and allow second children in the case of "practical difficulties" (such as cases in which the father is a disabled serviceman) or when both parents are single children,[21] and some provinces had other exemptions worked into their policies as well. In most areas, families were allowed to apply to have a second child if their first-born is a daughter.[18][22] Furthermore, families with children with disabilities have different policies and families whose first child suffers from physical disability, mental illness, or intellectual disability were allowed to have more children.[23] However, second children were sometimes subject to birth spacing (usually 3 or 4 years). Children born in overseas countries were not counted under the policy if they do not obtain Chinese citizenship. Chinese citizens returning from abroad were allowed to have a second child.[24] Sichuan province allowed exemptions for couples of certain backgrounds.[25] By one estimate there were at least 22 ways in which parents could qualify for exceptions to the law towards the end of the one-child policy's existence.[19] As of 2007, only 35.9% of the population were subject to a strict one-child limit. 52.9% were permitted to have a second child if their first was a daughter; 9.6% of Chinese couples were permitted two children regardless of their gender; and 1.6% — mainly Tibetans — had no limit at all.[26]

The Danshan, Sichuan Province Nongchang Village people Public Affairs Bulletin Board in September 2005 noted that RMB 25,000 in social compensation fees were owed in 2005. Thus far 11,500 RMB had been collected, so another 13,500 RMB had to be collected.

Following the 2008 Sichuan earthquake, a new exception to the regulations was announced in Sichuan province for parents who had lost children in the earthquake.[27][28] Similar exceptions had previously been made for parents of severely disabled or deceased children.[29] People have also tried to evade the policy by giving birth to a second child in Hong Kong, but at least for Guangdong residents, the one-child policy was also enforced if the birth was given in Hong Kong or abroad.[30]

In accordance with China's affirmative action policies towards ethnic minorities, all non-Han ethnic groups are subjected to different laws and were usually allowed to have two children in urban areas, and three or four in rural areas. Han Chinese living in rural towns were also permitted to have two children.[31] Because of couples such as these, as well as urban couples who simply pay a fine (or "social maintenance fee") to have more children,[32] the overall fertility rate of mainland China is close to 1.4 children per

woman.[33]

The Family Planning Policy was enforced through a financial penalty in the form of the "social child-raising fee", sometimes called a "family planning fine" in the West, which was collected as a fraction of either the annual disposable income of city dwellers or of the annual cash income of peasants, in the year of the child's birth.[34] For instance, in Guangdong, the fee is between 3 and 6 annual incomes for incomes below the per capita income of the district, plus 1 to 2 times the annual income exceeding the average. Both members of the couple need to pay the fine.[35]

In 2013, Deputy Director Wang Peian of the National Health and Family Planning Commission said that "China's population will not grow substantially in the short term".[36] A survey by the commission found that only about half of eligible couples wish to have two children, mostly because of the cost of living impact of a second child.[37]

In November 2013, following the Third Plenum of the 18th Central Committee of the Chinese Communist Party, China announced the decision to relax the one-child policy. Under the new policy, families could have two children if one parent, rather than both parents, was an only child.[38][39] This mainly applied to urban couples, since there were very few rural only children due to long-standing exceptions to the policy for rural couples.[40] The coastal province of Zhejiang, one of China's most affluent, became the first area to implement this "relaxed policy" in January 2014.[41] The relaxed policy has been implemented in 29 out of the 31 provinces, with the exceptions of Xinjiang and Tibet. Under this policy, approximately 11 million couples in China are allowed to have a second child; however, only "nearly one million" couples applied to have a second child in 2014,[42] less than half the expected number of 2 million per year.[43] By May 2014, 241,000 out of 271,000 applications had been approved. Officials of China's National Health and Family Planning Commission claimed that this outcome was expected, and that "second-child policy" would continue progressing with a good start.[44]

Abolition

See also: Two-child policy § China

In October 2015, the Chinese news agency Xinhua announced plans of the government to abolish the one-child policy, now allowing all families to have two children, citing from a communiqué issued by the Communist Party "to improve the balanced development of population" — an apparent reference to the country's female-to-male sex ratio — and to deal with an aging population according to the Canadian Broadcasting Corporation.[*45][*46][*12][*47][*48][*49][*50][*51] The new law is effective from 1 January 2016 after it was passed in the standing committee of the National People's Congress on 27 December 2015.[*52][*53]

The rationale for the abolition is summarized by former *Wall Street Journal* reporter Mei Fong: "The reason China is doing this right now is because they have too many men, too many old people, and too few young people. They have this huge crushing demographic crisis as a result of the one-child policy. And if people don't start having more children, they're going to have a vastly diminished workforce to support a huge aging population." [*54] China's ratio is about five working adults to one retiree; the huge retiree community must be supported, and that will dampen future growth, according to Fong.

Since the citizens of China are living longer and having fewer children, the growth of the population imbalance is expected to continue, as reported by the Canadian Broadcasting Corporation which referred to a United Nations projections forecast that China will lose 67 million working-age people by 2030, while simultaneously doubling the number of elderly. That could put immense pressure on the economy and government resources." [*12] The longer term outlook is also pessimistic, based on an estimate by the Chinese Academy of Social Sciences, revealed by Cai Fang, deputy director. "By 2050, one-third of the country will be aged 60 years or older, and there will be fewer workers supporting each retired person." [*55]

Although many critics of China's reproductive restrictions approve of the policy's abolition, some say that the move to the two-child policy will not end forced sterilizations, forced abortions, or government control over birth permits.[*56] Others also state that the abolition is not a sign of the relaxation of authoritarian control in China. A reporter for CNN said, "It was not a sign that the party will suddenly start respecting personal freedoms more than it has in the past. No, this is a case of the party adjusting policy to conditions. ...The new policy, raising the limit to two children per couple, preserves the state's role." [*57][*58] The abolition may not achieve a significant benefit, as the Canadian Broadcasting Corporation analysis indicates: "Repealing the one-child policy may not spur a huge baby boom, however, in part because fertility rates are believed to be declining even without the policy's enforcement. Previous easings of the one-child policy have spurred fewer births than expected, and many people among China's younger generations see smaller family sizes as ideal." [*12] The CNN reporter adds that China's new prosperity is also a factor in the declining[*55] birth rate, saying, "Couples naturally decide to have fewer children as they move from the fields into the cities, become more educated, and when women establish careers outside the home." [*57]

6.3.2 Administration

The one-child policy was managed by the National Population and Family Planning Commission under the central government since 1981. The Ministry of Health of the People's Republic of China and the National Health and Family Planning Commission were made defunct and a new single agency National Health and Family Planning Commission took over national health and family planning policies in 2013. The agency reports to the State Council.

The policy was enforced at the provincial level through fines that were imposed based on the income of the family and other factors. "Population and Family Planning Commissions" existed at every level of government to raise awareness and carry out registration and inspection work.[*59]

6.3.3 Effects

Births averted

The Chinese government says that 400 million births were prevented, though this statistic was not independently verified. In *Newsweek*, for example, the cause/effect was disputed, with "...some demographers claim[ing] China's population growth would have flattened out even without it—the draconian rule left emotional, social and economic scars the country and its citizens will be dealing with for years." [*60]

Continuation of Demographic Transition stage three

Further information: Demographics of China
The fertility rate in China continued its fall from 2.8 births

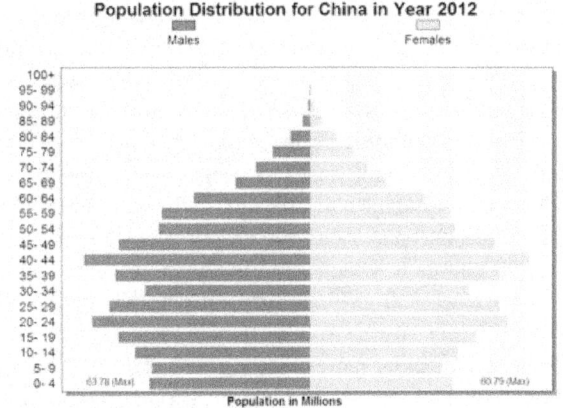

The progression of China's population pyramid, *International Futures.*

per woman in 1979 (already a sharp reduction from more

than five births per woman in the early 1970s) to 1.5 in 2010.*[61] This is similar to demographic transition seen in Thailand, Indian states of Kerala, Tamil Nadu which have undergone similar changes in fertility rates without a one child policy.*[62]*[63]*[61]*[64]

Disparity in sex ratio at birth

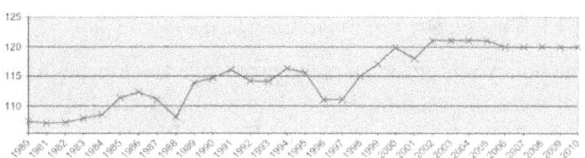

The sex ratio at birth in mainland China, males per 100 females, 1980–2010.

For more details on this topic, see Missing women of China.

The sex ratio at birth (between male and female births) in mainland China reached 117:100 and remained steady between 2000 and 2013, substantially higher than the natural baseline, which ranges between 103:100 and 107:100. It had risen from 108:100 in 1981—at the boundary of the natural baseline—to 111:100 in 1990.*[65] According to a report by the National Population and Family Planning Commission, there will be 30 million more men than women in 2020, potentially leading to social instability, and courtship-motivated emigration.*[66]

The disparity in the gender ratio at birth increases dramatically after the first birth, for which the ratios remained steadily within the natural baseline over the 20 year interval between 1980 and 1999. Thus, a large majority of couples appear to accept the outcome of the first pregnancy, whether it is a boy or a girl. If the first child is a girl, and they are able to have a second child, then a couple may take extraordinary steps to assure that the second child is a boy. If a couple already has two or more boys, the sex ratio of higher parity births swings decidedly in a feminine direction. This demographic evidence indicates that while families highly value having male offspring, a secondary norm of having a girl or having some balance in the sexes of children often comes into play.Zeng 1993 reported a study based on the 1990 census in which they found sex ratios of just 65 or 70 boys per 100 girls for births in families that already had two or more boys.*[67] A study by Anderson & Silver 1995 found a similar pattern among both Han and non-Han nationalities in Xinjiang Province: a strong preference for girls in high parity births in families that had already borne two or more boys.*[68] This tendency to favour girls in high parity births to couples who had already borne sons was later also noted by Coale, who suggested as well that once a couple had achieved its goal for the number of males, it was

also much more likely to engage in "stopping behavior", i.e., to stop having more children.*[69]

The long-term disparity has led to a significant gender imbalance or skewing of the sex ratio. As reported by the Canadian Broadcasting Corporation, China has between 32 million and 36 million more males than would be expected naturally, and this has led to social problems. "Because of a traditional preference for baby boys over girls, the one-child policy is often cited as the cause of China's skewed sex ratio... Even the government acknowledges the problem and has expressed concern about the tens of millions of young men who won't be able to find brides and may turn to kidnapping women, sex trafficking, other forms of crime or social unrest." *[12] The situation will not improve in the near future. According to the Chinese Academy of Social Sciences, there will be 24 million more men than women of marriageable age by 2020.*[70]

Adoption

A roadside sign in rural Sichuan: "It is forbidden to discriminate against, mistreat or abandon baby girls."

The one-child policy of China made it more expensive for parents with children to adopt, which may have had an effect upon the numbers of children living in state-sponsored orphanages. However, in the 1980s and early 1990s, poor care and high mortality rates in some state institutions generated intense international pressure for reform.*[71]*[72]

In the 1980s, adoptions accounted for half of the so-called "missing girls" .*[73] Through the 1980s, as the one-child policy came into force, parents who desired a son but had a daughter often failed to report or delayed reporting female births to the authorities. Some parents may have offered up their daughters for formal or informal adoption. A majority of children who went through formal adoption in China in the later 1980s were girls, and the proportion who were girls increased over time.*[73]

In an interview with National Public Radio on October 30, 2015, Adam Pertman,[*][74] president and CEO of the National Center on Adoption and Permanency, indicated that many young girls were adopted by citizens of other countries, particularly the United States, a trend which has been declining for some years. "The infant girls of yesteryear have not been available, if you will, for five, seven years. China has been [⋯] trying to keep the girls within the country [⋯]. And the consequence is that, today, rather than those young girls who used to be available – primarily girls – today, it's older children, children with special needs, children in sibling groups. It's very, very different."[*][75]

A government sign in Tangshan Township: "For a prosperous, powerful nation and a happy family, please practice family planning."

Twins

Since there are no penalties for multiple births, it is believed that an increasing number of couples are turning to fertility medicines to induce the conception of twins. According to a 2006 *China Daily* report, the number of twins born per year was estimated to have doubled.[*][76]

Quality of life for women

The one-child policy has played a major role in improving the quality of life for women in China. For thousands of years, girls have held a lower status in Chinese households. However, the one-child policy's limit on the number of children has prompted parents of women to start investing money in their well-being. As a result of being an only child, women have increased opportunity to receive an education, and support to get better jobs.[*][77]

Healthcare improvements

It is reported that the focus of China on population control helps provide a better health service for women and a reduction in the risks of death and injury associated with pregnancy. At family planning offices, women receive free contraception and pre-natal classes that contributed to the policy's success in two respects. First, the average Chinese household expends fewer resources, both in terms of time and money, on children, which gives many Chinese people more money with which to invest. Second, since Chinese adults can no longer rely on children to care for them in their old age, there is an impetus to save money for the future.[*][78]

"Four-two-one" problem

As the first generation of law-enforced only-children came of age for becoming parents themselves, one adult child was left with having to provide support for his or her two parents and four grandparents.[*][79][*][80] Called the "4-2-1 Problem", this leaves the older generations with increased chances of dependency on retirement funds or charity in order to receive support. If personal savings, pensions, or state welfare fail, most senior citizens would be left entirely dependent upon their very small family or neighbours for assistance. If, for any reason, the single child is unable to care for their older adult relatives, the oldest generations would face a lack of resources and necessities. In response to such an issue, all provinces have decided that couples are allowed to have two children if both parents were only children themselves: By 2007, all provinces in the nation except Henan had adopted this new policy;[*][81][*][82] Henan followed in 2011.[*][83]

Unregistered children

Further information: Heihaizi

Heihaizi (Chinese: 黑孩子; pinyin: *hēiháizi*) or "black child" is a term applied in China. The term denotes children born outside the One child policy, or generally children who are not registered in the Chinese national household registration system.

Being excluded from the family register (in effect, a birth certificate), they do not legally exist and as a result cannot access most public services, such as education and health care, and do not receive protection under the law.[*][84][*][85][*][86]

Potential social problems

Some parents may over-indulge their only child. The media referred to the indulged children in one-child families as "little emperors". Since the 1990s, some people have worried that this will result in a higher tendency toward poor social communication and cooperation skills amongst the new generation, as they have no siblings at home. No social studies have investigated the ratio of these over-indulged children and to what extent they are indulged. With the first generation of children born under the policy (which initially became a requirement for most couples with first children born starting in 1979 and extending into the 1980s) reaching adulthood, such worries were reduced.[*][87]

However, the "little emperor syndrome" and additional expressions, describing the generation of Chinese singletons are very abundant in the Chinese media, Chinese academia and popular discussions. Being over-indulged, lacking self-discipline and having no adaptive capabilities are traits that are highly associated with Chinese singletons.[*][88]

Some 30 delegates called on the government in the Chinese People's Political Consultative Conference in March 2007 to abolish the one-child rule, citing "social problems and personality disorders in young people". One statement read, "It is not healthy for children to play only with their parents and be spoiled by them: it is not right to limit the number to two children per family, either." [*][89] The proposal was prepared by Ye Tingfang, a professor at the Chinese Academy of Social Sciences, who suggested that the government at least restore the previous rule that allowed couples to have up to two children. According to a scholar, "The one-child limit is too extreme. It violates nature's law. And in the long run, this will lead to mother nature's revenge." [*][89][*][90]

Birth tourism

Reports surfaced of Chinese women giving birth to their second child overseas, a practice known as birth tourism. Many went to Hong Kong, which is exempt from the one-child policy. Likewise, a Hong Kong passport differs from China mainland passport by providing additional advantages. Recently though, the Hong Kong government has drastically reduced the quota of births set for non-local women in public hospitals. As a result, fees for delivering babies there have surged. As further admission cuts or a total ban on non-local births in Hong Kong are being considered, mainland agencies that arrange for expectant mothers to give birth overseas are predicting a surge in those going to North America.[*][91]

As the United States practises birthright citizenship, children born in the US will be US citizens. The closest option (from China) is Saipan in the Northern Mariana Islands, a US dependency in the western Pacific Ocean that allows Chinese visitors without visa restrictions. The island is currently experiencing an upswing in Chinese births. This option is used by relatively affluent Chinese who often have secondary motives as well, wishing their children to be able to leave mainland China when they grow older or bring their parents to the US. Canada is less achievable as their government denies many visa requests.[*][92][*][93]

6.3.4   Criticism

The policy is controversial outside China for many reasons, including accusations of human rights abuses in the implementation of the policy, as well as concerns about negative social consequences.[*][94]

Statement of the effect of the policy on birth reduction

The Chinese government, quoting Zhai Zhenwu, director of Renmin University's School of Sociology and Population in Beijing, estimates that 400 million births were prevented by the one-child policy as of 2011, while some demographers challenge that number, putting the figure at perhaps half that level, according to CNN.[*][95] Zhai clarified that the 400 million estimate referred not just to the one-child policy, but includes births prevented by predecessor policies implemented one decade before, stating that "there are many different numbers out there but it doesn't change the basic fact that the policy prevented a really large number of births." [*][96]

This claim is disputed by Wang Feng, director of the Brookings-Tsinghua Center for Public Policy, and Cai Yong from the Carolina Population Center at University of North Carolina Chapel Hill[*][96] Wang claims that "Thailand and China have had almost identical fertility trajectories since the mid 1980s," and "Thailand does not have a one-child policy." [*][96] China's Health Ministry has also disclosed that at least 336 million abortions were performed on account of the policy.[*][97]

According to a report by the US Embassy, scholarship published by Chinese scholars and their presentations at the October 1997 Beijing conference of the International Union for the Scientific Study of Population seemed to suggest that market-based incentives or increasing voluntariness is not morally better but that it is in the end more effective.[*][98] In 1988, Zeng Yi and Professor T. Paul Schultz of Yale University discussed the effect of the transformation to the market on Chinese fertility, arguing that the introduction of the contract responsibility system in agriculture during the early 1980s weakened family planning controls during that period.[*][99] Zeng contended that the "big cooking pot" system of the People's Communes had insulated people from the costs of having many children. By the late 1980s, economic costs and incentives created by the contract system were already reducing the number of children farmers wanted.

A long-term experiment in a county in Shanxi Province where the family planning law was suspended, that suggested that families would not have many more children even if the law were abolished.*[19] A 2003 review of the policy-making process behind the adoption of the one-child policy shows that less intrusive options, including those that emphasized delay and spacing of births, were known but not fully considered by China's political leaders.*[100]

Unequal enforcement

Corrupted government officials and especially wealthy individuals have often been able to violate the policy in spite of fines.*[101] Filmmaker Zhang Yimou had three children and was subsequently fined 7.48 million yuan ($1.2 million).*[102] For example, between 2000 and 2005, as many as 1,968 officials in central China's Hunan province were found to be violating the policy, according to the provincial family planning commission; also exposed by the commission were 21 national and local lawmakers, 24 political advisors, 112 entrepreneurs and 6 senior intellectuals.*[101]

Some of the offending officials did not face penalties,*[101] although the government did respond by raising fines and calling on local officials to "expose the celebrities and high-income people who violate the family planning policy and have more than one child."*[101] Also, people who lived in the rural areas of China were allowed to have two children without punishment, although the family is required to wait a couple of years before having another child.*[103]

Human rights violations

Further information: Human rights in China

The one-child policy has been challenged for violating a human right to determine the size of one's own family. According to a 1968 proclamation of the International Conference on Human Rights, "Parents have a basic human right to determine freely and responsibly the number and the spacing of their children." *[104]*[105]

According to the UK newspaper *The Daily Telegraph*, a quota of 20,000 abortions and sterilizations was set for Huaiji County in Guangdong Province in one year due to reported disregard of the one-child policy. According to the article local officials were being pressured into purchasing portable ultrasound devices to identify abortion candidates

in remote villages. The article also reported that women as far along as 8.5 months pregnant were forced to abort, usually by an injection of saline solution.*[106] A 1993 book by social scientist Steven W. Mosher reported that women in their ninth month of pregnancy, or already in labour, were having their children killed whilst in the birth canal or immediately after birth.*[107]

According to a 2005 news report by Australian Broadcasting Corporation correspondent John Taylor, China outlawed the use of physical force to make a woman submit to an abortion or sterilization in 2002 but ineffectively enforces the measure.*[108] In 2012, Feng Jianmei, a villager from central China's Shaanxi province was forced into an abortion by local officials after her family refused to pay the fine for having a second child. Chinese authorities have since apologized and two officials were fired, while five others were sanctioned.*[109]

In the past China promoted eugenics as part of its population planning policies, but the government has backed away from such policies, as evidenced by China's ratification of the Convention on the Rights of Persons with Disabilities, which compels the nation to significantly reform its genetic testing laws.*[110] Recent research has also emphasized the necessity of understanding a myriad of complex social relations that affect the meaning of informed consent in China.*[111] Furthermore, in 2003, China revised its marriage registration regulations and couples no longer have to submit to a pre-marital physical or genetic examination before being granted a marriage license.*[112]

The United Nations Population Fund's (UNFPA) support for family planning in China, which has been associated with the One-Child policy in the United States, led the United States Congress to pull out of the UNFPA during the Reagan administration,*[113] and again under George W. Bush's presidency, citing human rights abuses*[114] and stating that the right to "found a family" was protected under the Preamble in the Universal Declaration of Human Rights.*[115] President Obama resumed U.S. government financial support for the UNFPA shortly after taking office in 2009, intending to "work collaboratively to reduce poverty, improve the health of women and children, prevent HIV/AIDS and provide family planning assistance to women in 154 countries" .*[116]*[117]

Effect on infanticide rates

Sex-selected abortion, abandonment, and infanticide are illegal in China. Nevertheless, the United States Department of State,*[118] the Parliament of the United Kingdom,*[119] and the human rights organization Amnesty International*[120] have all declared that infanticide still exists.*[121]*[122]*[123] A writer for the Georgetown Journal of International Affairs wrote, "The 'one-child' policy has also led to what Amartya Sen first called 'Missing Women', or the 100 million girls 'missing' from the populations of China (and other developing countries) as a result of female infanticide, abandonment, and neglect".*[124]

The Canadian Broadcasting Corporation offered the following summary as to the long term effects of sex-selective abortion and abandonment of female infants: "Multiple research studies have also found that sex-selective abortion —where a woman undergoes an ultrasound to determine the sex of her baby, and then aborts it if it's a girl —was widespread for years, particularly for second or subsequent children. Millions of female fetuses have been aborted since the 1970s. China outlawed sex selective abortions in 2005, but the law is tough to enforce because of the difficulty of proving why a couple decided to have an abortion. The abandonment, and killing, of baby girls has also been reported, though recent research studies say it has become rare, in part due to strict criminal prohibitions." *[12]

Anthropologist G. William Skinner at the University of California, Davis and Chinese researcher Yuan Jianhua have claimed that infanticide was fairly common in China before the 1990s.*[125]

6.3.5 In popular culture

- Ball, David (2002). *China Run*. Simon & Schuster. ISBN 0-74322743-3. A novel about an American woman who travels to China to adopt an orphan of the one-child policy, only to find herself a fugitive when the Chinese government informs her that she has been given "the wrong baby."

- The prevention of a state imposed abortion during labor to conform with the one child policy was a key plot point in Tom Clancy's novel *The Bear and the Dragon*.

- The difficulties of implementing the one-child policy are dramatized in Mo Yan's novel *Frog* (2009; English translation by Howard Goldblatt, 2015).

- Avoiding the family-planning enforcers is at the heart of Ma Jian's novel *The Dark Road* (translated by Flora Drew, 2013).

- Novelist Lu Min writes about her own family's experience with the One Child Policy in her essay *A Second Pregnancy, 1980* (translated by Helen Wang, 2015).*[126]

- Xue, Xinran (2015). *Buy Me the Sky*. Rider (imprint). ISBN 978-1-8460-4471-7. Tells the stories of the children brought up under China's One-child policy and the effect that has had on their lives, families and ability to deal with life's challenges.

6.3.6 See also

- Shidu (parents), denoting the loss of an only child
- Two-child policy
- *The Dying Rooms*

General:

- Abortion in China
- Demographics of China
- Human population control
- List of countries and dependencies by population
- Human overpopulation

6.3.7 References

[1] "The Chinese Celebrate Their Roaring Economy, As They Struggle With Its Costs". Retrieved 9 December 2015.

[2] Bergaglio, Maristella. "Population Growth in China: The Basic Characteristics of China's Demographic Transition" (PDF). *Global geografia* (IT).

[3] "World Development Indicators". *Google Public Data Explorer*. World Bank. 2009-07-01. Retrieved 2013-10-04.

[4] Mann, Jim (1992-06-07). "The Physics of Revenge: When Dr. Lu Gang's American Dream Died, Six People Died With It". *The Los Angeles Times Magazine*. Retrieved July 14, 2012.

[5] Potts, M. (19 August 2006). "China's one child policy". *BMJ* **333** (7564): 361–62. doi:10.1136/bmj.38938.412593.80. PMC: 1550444. PMID 16916810.

[6] "Total population, CBR, CDR, NIR and TFR of China (1949–2000)". *China Daily*. Retrieved 2013-10-04.

[7] Zubrin, Robert (2012). *Radical Environmentalists, Criminal Pseudo-Scientists, and the Fatal Cult of Antihumanism*. The New Atlantis. 2646. ISBN 978-1-59403476-3.

[8] *Family Planning in China*, Embassy of the People's Republic of China in Lithuania; Information Office of the State Council of the People's Republic of China, August 1995, Section III paragraph 2, retrieved 27 October 2014

[9] Zhu, W X (1 June 2003). "The One Child Family Policy". *Archives of Disease in Childhood* **88** (6): 463–64. doi:10.1136/adc.88.6.463. PMC: 1763112. PMID 12765905.

[10] "East and Southeast Asia: China". *CIA World Factbook*.

[11] Coale, Ansley J. (March 1981). "Population Trends, Population Policy, and Population Studies in China" (PDF). *Population and Development Review* **7** (1): 85. doi:10.2307/1972766. Coale shows detailed birth and death data up to 1979, and gives a cultural background to the famine in 1959–61.

[12] *Five things to know about China's one-child policy*, CA: CBC.

[13] da Silva, Pascal Rocha (2006). "La politique de l'enfant unique en République populaire de Chine" [The politics of one child in the People's Republic of China] (PDF) (in French). University of Geneva: 22–28.

[14] "Experts challenge China's 1-child population claim". *Boston.com*. 2011-10-27.

[15] Fong, Vanessa L. (2004). *Only Hope: Coming of Age Under China's One-Child Policy*. Stanford University Press. p. 179. ISBN 978-0-80475330-2.

[16] "Regulations on Family Planning of Henan Province". Henan Daily. 5 April 2000. Archived from the original on 9 July 2008. Retrieved 29 October 2008.

[17] 国务院专家: 建议全面放开二胎, *A finance* (in Chinese) (CN: Yaolan), 6 July 2012, Article 13, retrieved 17 July 2012.

[18] "Most people free to have more child". *China Daily*. 2007-07-11. Retrieved 2009-07-31.

[19] Wong, Edward (July 22, 2012). "Reports of Forced Abortions Fuel Push to End Chinese Law". *The New York Times*. Retrieved July 23, 2012.

[20] "Status of Population and Family Planning Program in China by Province". Economic and Social Commission for Asia and the Pacific. Archived from the original on 30 March 2012.

[21] Scheuer, James (4 January 1987). "America, the U.N. and China's Family Planning (Opinion)". *The New York Times*. Retrieved 27 October 2008.

[22] Hu, Huiting (18 October 2002). "Family Planning Law and China's Birth Control Situation". *China Daily*. Retrieved 2 March 2009.

[23] "China's Only Child". *NOVA*. 14 February 1984. PBS. Retrieved 13 October 2009.

[24] Qiang, Guo (2006-12-28). "Are the rich challenging family planning policy?". *China Daily*.

[25] *29th session of the standing committee of the 8th People's Congress of Sichuan Province* (rev ed.), United Nations Economic and Social Commission for Asia and the Pacific, 17 October 1997, Articles 11–13, archived from the original on 6 July 2008, retrieved 31 October 2008

[26] Callick, Rowan (24 January 2007). "China relaxes its one-child policy". *The Australian*.

[27] Jacobs, Andrew Jacobs (27 May 2008). "One-Child Policy Lifted for Quake Victims' Parents". *The New York Times*. Retrieved 28 May 2008.

[28] "Baby offer for earthquake parents". BBC. Retrieved 31 October 2008.

[29] "China Amends Child Policy for Some Quake Victims". *Morning Edition*. NPR.

[30] Tan, Kenneth (2012-02-09). "Hong Kong to issue blanket ban on mothers from the mainland?". Shanghaiist. Retrieved 2013-10-04.

[31] Yardley, Jim (11 May 2008). "China Sticking With One-Child Policy". *The New York Times*. Retrieved 20 November 2008.

[32] "New rich challenge family planning policy". *Xinhua*. 2015-12-14.

[33] "The most surprising demographic crisis". *The Economist*. 5 May 2011. Retrieved 25 February 2013.

[34] Summary of Family Planning notice on how FP fines are collected

[35] "Heavy Fine for Violators of One-Child Policy". CN. Retrieved 2013-10-04.

[36] Burkitt, Laurie (2013-11-17), "China to Move Slowly on One-Child Law Reform", *The Wall Street Journal* (online ed.), retrieved 2013-12-05.

[37] Levin, Dan (25 February 2014). "Many in China Can Now Have a Second Child, but Say No". *The New York Times*. Retrieved 26 February 2014.

[38] *China reforms: One-child policy to be relaxed*, UK: BBC, 2013-11-15, retrieved 2013-12-05.

[39] "Why is China relaxing its one-child policy?". *The Economist*. The Economist. 27 January 2015. Retrieved 27 January 2015.

[40] "Xinhua Insight: Heated discussion over loosening of one-child policy". *Xinhua net*.

[41] "Eastern Chinese province first to ease one-child policy". *Reuters*. 17 January 2014.

[42] "1 mln Chinese couples apply to have second child". *China daily*.

[43] *China daily*, Feb 2014.

[44] Wang, Yamei (2014). "11 million couples qualify for a second child". *Xinhua News*. Retrieved December 10, 2014.

[45] "China to abolish decades-old one-child policy". Al Jazeera English. 29 October 2015. Retrieved 30 October 2015.

[46] Jiang, Steven; Hanna, Jason (29 October 2015). "China says it will end one-child policy". CNN. Retrieved 29 October 2015.

[47] "Beschluss der Kommunistischen Partei: China beendet Ein-Kind-Politik" (in German). DE: Tagesschau. 29 October 2015. Retrieved 29 October 2015.

[48] "China to end one-child policy and allow two". *BBC News*.

[49] "China to allow two children for all couples". Xinhua. 29 October 2015.

[50] Phillips, Tom. "China ends one-child policy after 35 years". *The Guardian*.

[51] "The 'model' example of China's one child policy". *BBC News*.

[52] "Top legislature amends law to allow all couples to have two children". Xinhua News Agency. 27 December 2015.

[53] "China formally abolishes decades-old one-child policy". International Business Times. 27 December 2015.

[54] Fong, Mei (2015-10-15), "China one-child policy", *National Geographic*.

[55] *China daily*, Dec 2014.

[56] "China ends one-child policy —but critics warn new two-child policy won't end forced abortions". The Raw Story. 29 October 2015. Retrieved 29 October 2015.

[57] Ghitis (2015-10-29), *China: one-child policy*, CNN.

[58] http://www.ejinsight.com/20151105-china-one-child-calamity/

[59] Dewey, Arthur E (16 December 2004). "One-Child Policy in China". Senior State Department.

[60] "Its one-child policy lifted, China becoming World's largest old-age home", *Newsweek*

[61] Feng, Wang; Yong, Cai; Gu, Baochang (2012). "Population, Policy, and Politics: How Will History Judge China's One-Child Policy?" (PDF). *Population and Development Review* **38**: 115–29. doi:10.1111/j.1728-4457.2013.00555.x.

[62] Sen, Amartya. "Population Policy: Authoritarianism versus Cooperation" (PDF). BR: Universidade de Campinas.

[63] Sen, Amartya (Jun 2012). "Population: Delusion and Reality" (PDF). Richard R Guzmán.

[64] Cai, Yong (Sep 2010). "China's Below-Replacement Fertility: Government Policy or Socioeconomic Development?" (PDF). *Population and Development Review* **36** (3): 419–40. doi:10.1111/j.1728-4457.2010.00341.x.

[65] Wei, Chen (2005). "Sex Ratios at Birth in China" (PDF). Archived from the original (PDF) on 18 July 2006. Retrieved 2 March 2009.

[66] "Chinese facing shortage of wives". BBC. 2007-01-12. Retrieved 2007-01-12.

[67] Zeng, Yi; et al (1993), "Causes and Implications of the Recent Increase in the Reported Sex Ratio at Birth in China", *Population and Development Review* **19** (June): 283–302, doi:10.2307/2938438 Cite uses deprecated parameter |coauthors= (help).

[68] Anderson, Barbara A; Silver, Brian D (1995), "Ethnic Differences in Fertility and Sex Ratios at Birth in China: Evidence from Xinjiang", *Population Studies* **49** (July): 211–26, doi:10.1080/0032472031000148476.

[69] Coale, Ansley J (1996). "Five Decades of Missing Females in China". *Proceedings of the American Philosophical Society* **140** (4): 421–50. doi:10.2307/2061752. JSTOR 987286. PMID 7828766.

[70] "Online dating a path to marriage for young, busy Chinese", *Beijing today*, Oct 2015.

[71] *Death by Default: A Policy of Fatal Neglect in China's State Orphanages*. New York: Human Rights Watch/Asia. 1996. ISBN 1-56432-163-0.

[72] "Chinese Orphanages: A Follow-up" (PDF). Human Rights Watch/Asia. March 1996.

[73] Johansson, Sten; Nygren, Olga (1991). "The missing girls of China: a new demographic account". *Population and Development Review* (Population Council) **17** (1): 35–51. doi:10.2307/1972351. JSTOR 1972351.

[74] *Adam Pertman*, National center on adoption & permanency.

[75] *How China's one-child policy transformed US attitudes on adoption*, NPR, 2015-10-30.

[76] "China: Drug bid to beat child ban". *China Daily*. Associated Press. 14 February 2006. Retrieved 11 November 2008.

[77] "How China's one-child policy overhauled the status and prospects of girls like me". *The Daily Telegraph*. Retrieved 18 February 2016.

[78] Naughton, Barry (2007). *The Chinese Economy: Transitions and Growth*. Cambridge, Mass.: MIT Press. ISBN 978-0262640640.

[79] 李雯 [Li Wen] (5 April 2008). "四二一" 家庭，路在何方？ ['Four-two-one families', where is the road going?] (in Chinese). 云南日报网 [Yunnan Daily Online]. Archived from the original on 18 March 2011. Retrieved 31 January 2011.

[80] 四二一" 家庭真的是问题吗？ [Are 'four-two-one' families really a problem?] (in Chinese). 中国人口学会网 [China Population Association Online]. 10 October 2010. Archived from the original on 2011-07-07. Retrieved 31 January 2011.

[81] "Rethinking China's one-child policy". CBC. October 28, 2009. Retrieved June 11, 2010.

[82] 计生委新闻发言人:11% 以上人口可生两个孩子[Spokesperson of the one-child policy committee: 11% or more of the population may have two children] (in Chinese). Sina. 10 July 2007. Retrieved 7 November 2008.

[83] "China's most populous province amends family-planning policy". *People's Daily Online*. 2011-11-25.

[84] 黒核子～一人っ子政策の大失敗[Black Children - The Failure of One Child Policy] (in Japanese). Retrieved 10 July 2010.

[85] "One Child Policy - Laogai Research Foundation (LRF)". *Laogai Research Foundation*. Retrieved 13 July 2010.

[86] Li, Shuzhuo; Zhang, Yexia; Feldman, Marcus W (2010). "Birth Registration in China: Practices, Problems and Policies". *Population research and policy review* 29 (3): 297–317. doi:10.1007/s11113-009-9141-x. PMC: 2990197. PMID 21113384.

[87] Deane, Daniela (July 26, 1992). "The Little Emperors". *The Los Angeles Times*. p. 16.

[88] "Chinese Singletons – Basic 'Spoiled' Related Vocabulary". *Thinking Chinese*. November 11, 2010.

[89] "Consultative Conference: 'The government must end the one-child rule'". IT: AsiaNews. 2007-03-16.

[90] "Advisors say it's time to change one-child policy". Shanghai Daily. 2007-03-15.

[91] "談天説地". review33. Archived from the original on October 4, 2013. Retrieved 2013-10-04.

[92] Eugenio, Haidée V. "Birth tourism on the upswing". *Saipan Tribune*.

[93] Eugenio, Haidée V. "Many Chinese giving birth in CNMI trying to get around one child policy". *Saipan Tribune*.

[94] Hvistendahl, Mara (17 September 2010). "Has China Outgrown The One-Child Policy?". *Science* 329 (5998): 1458–61. doi:10.1126/science.329.5998.1458. PMID 20847244.

[95] Some demographers challenge that number, putting the figure at perhaps half that level.

[96] "Experts challenge China's 1-child population claim".

[97] "336 million abortions under China's one-child policy". *Telegraph.co.uk*. 15 March 2013.

[98] "PRC Family Planning: The Market Weakens Controls But Encourages Voluntary Limits". U.S. Embassy in Beijing. June 1988.

[99] PRC journal *Social Sciences in China* [Zhongguo , January 1988]

[100] Greenhalgh, Susan (2003). "Science, Modernity, and the Making of China's One-Child Policy" (PDF). *Population and Development Review* 29 (June): 163–196. doi:10.1111/j.1728-4457.2003.00163.x.

[101] "Over 1,900 officials breach birth policy in C. China". *Xinhua*. 8 July 2007. Archived from the original on 10 October 2008. Retrieved 11 November 2008. But heavy fines and exposures seemed to hardly stop the celebrities and rich people, as there are still many people, who can afford the heavy penalties, insist on having multiple kids, the Hunan commission spokesman said...Three officials... who were all found to have kept extramarital mistresses, were all convicted for charges such as embezzlement and taking bribes, but they were not punished for having more than one child.

[102] "China: Filmmaker Zhang Yimou fined $1M for breach of one-child policy - CNN.com". *CNN*. Retrieved 2016-01-03.

[103] chan, peggy (2005). *Cultures of the world China*. New York: Marshall Cavendish International.

[104] Freedman, Lynn P.; Isaacs, Stephen L. (Jan–Feb 1993). "Human Rights and Reproductive Choice" (PDF). *Studies in Family Planning* (Population Council) 24 (1): 18–30. doi:10.2307/2939211. JSTOR 2939211. PMID 8475521. Retrieved 2007-12-08.

[105] "Proclamation of Teheran". International Conference on Human Rights. 1968. Archived from the original on 2007-10-17. Retrieved 2007-11-08.

[106] McElroy, Damien (2001-04-08). "Chinese region 'must conduct 20,000 abortions'". *The Telegraph* (London).

[107] Mosher, Steven W. (July 1993). *A Mother's Ordeal*. Harcourt. ISBN 0-15-162662-6.

[108] Taylor, John (2005-02-08). "China – One Child Policy". Australian Broadcasting Corporation. Retrieved 2008-07-01.

[109] "Father in forced abortion case wants charges filed". *My Way News*. Associated Press. 2012-07-06.

[110] (subscription required) "Implications of China's Ratification of the United Nations Convention on the Rights of Persons with Disabilities". *China: an International Journal*.

[111] Sleeboom-Faulkner, Margaret Elizabeth (1 June 2011). "Genetic testing, governance, and the family in the People's Republic of China". *Social Science & Medicine* 72 (11): 1802–9. doi:10.1016/j.socscimed.2010.03.052.

[112] "Marriage Law of the People's Republic of China" (PDF). *Australia: Refugee Review Tribunal*.

[113] Moore, Stephen (1999-05-09). "Don't Fund UNFPA Population Control". CATO Institute.

[114] McElroy, Damien (2002-02-03). "China is furious as Bush halts UN 'abortion' funds". *The Telegraph* (London).

[115] Siv, Sichan (2003-01-21). "United Nations Fund for Population Activities in China". U.S. Department of State. Archived from the original on 19 February 2003.

[116] "UNFPA Welcomes Restoration of U.S. Funding". *UNFPA News*. 29 January 2009. Archived from the original on 12 May 2013.

[117] Rizvi, Haider (March 12, 2009). "Obama Sets New Course at the U.N.". *IPS News*. Inter Press Agency.

[118] Associated Press. "US State Department position". Archived from the original on December 20, 2008.

[119] "Human Rights in China and Tibet". Parliament of the United Kingdom.

[120] Amnesty International. "Violence Against Women – an introduction to the campaign". Archived from the original on 9 October 2006.

[121] Mosher, Steve (1986). "Steve Mosher's China report". *The Interim*.

[122] "Case Study: Female Infanticide". *Gendercide Watch*. 2000. Archived from the original on May 26, 2013.

[123] "Infanticide Statistics: Infanticide in China". *All Girls Allowed*. 2010.

[124] Steffensen, Jennifer. "Georgetown Journal's Guide to the 'One-Child' Policy". Retrieved 2013-09-30.

[125] Lubman, Sarah (2000-03-15). "Experts Allege Infanticide In China — 'Missing' Girls Killed, Abandoned, Pair Say". *San Jose Mercury News* (CA).

[126] "A Second Pregnancy, 1980", *Paper republic*.

6.3.8 Further reading

- *Better 10 Graves Than One Extra Birth: China's Systemic Use of Coercion To Meet Population Quotas*. Washington, DC: Laogai Research Foundation. 2004. ISBN 1-931550-92-1.

- Goh, Esther C.L. (2011). "China's One-Child Policy and Multiple Caregiving: raising little suns in Xiamen" (PDF). *Journal of International and Global Studies* (New York: Routledge).

- Greenhalgh, Susan (2008). *Just One Child: Science and Policy in Deng's China* (illustrated ed.). University of California Press. ISBN 978-0-520-25339-1.

- Fong, Mei (2015). *One Child: The Past and Future of China's Most Radical Experiment*. Houghton Mifflin Harcourt. ISBN 0-544-27539-X.

6.3.9 External links

- Family Planning in China

- "Illegal births and legal abortions – the case of China". *Reprod Health* **2**: 5. 2005. doi:10.1186/1742-4755-2-5. PMC: 1215519. PMID 16095526.

6.4 Tan Kai

This is a Chinese name; the family name is 谭 (Tan).

Tan Kai (Chinese: 谭凯; born 1973) is a mainland Chinese computer technician and an environmental activist from Zhejiang province. He operated his own company, called Lanyi Computer Repair, and co-founded an environmental advocacy and monitoring NGO called Green Watch (绿色观察). He was convicted in May 2006 "illegally obtaining state secrets."

6.4.1 Environmental activism

Tan became interested in environmental issues following April 2005's violent struggles over pollution and corruption in the town of Huashui, Zhejiang, where many residents believe that releases of toxic substances from chemical plants into the water supply are destroying crops and causing birth defects.[*][1] Further riots in Dongyang, in Xinchang (over a pharmaceutical factory), and at a battery factory in Changxin, convinced Tan to set up an environmental monitoring group, which he did informally in the summer of that year, together with five other individuals: Mr. Lai Jinbiao, Mr. Gao Haibing, Mr. Wu Yuanming, Mr. Qi Huimin, and Mr. Yang Jianming.

Because in order to operate lawfully as a local organization China requires a large staff, an office, and a large sum of money, in October 2005 Tan opened an account at a branch of Bank of China in Hangzhou with the sum of 500 yuan. When all six members of Green Watch were detained and released on October 19, Tan was charged and kept in custody. Although Tan was ostensibly arrested and charged with "illegally obtaining state secrets" after performing a routine backup on a computer belonging to a member of the Zhejiang Communist Party committee, Green Watch was declared illegal and banned one month later. Tan's friend and fellow activist Lai Jinbiao believes Tan was held because his name was the one on the bank account. On November 15, 2005, the Zhejiang provincial government declared Green Watch an illegal organization.

6.4.2 Imprisonment

Tan was held incommunicado for nearly seven months until May 9, 2006. During this time, his father engaged two Beijing-based lawyers, Li Heping and Li Xiongbing, to defend Tan, but the Hangzhou Public Security Bureau denied permission to engage counsel because the case involved state secrets. The elder Tan persisted with another application, and Li was finally able to meet with Tan for one hour at the West Lake Detention Centre in Hangzhou.

6.4.3 Trial

Although Tan pleaded innocent and no evidence of any crime was presented (the person from whom the secrets were supposedly taken did not appear), he was convicted in a three-hour trial at the Xihu District Court in Hangzhou, on the morning of May 15, 2006, which was closed to the public. He was sentenced to 18 months in prison for "illegally obtaining state secrets" by the Hangzhou Municipal People's Intermediate Court on August 11, 2006. His lawyer Li Heping raised concerns about Tan's health condition in prison, as he suffers from a liver disease.

6.4.4 Release

Tan was released in 2007 following the serving of his full 18-month sentence.

6.4.5 See also

- Environment of China
- Wu Lihong

6.4.6 References

[1] The Epoch Times | Large Scale Riot Erupts in Huashui Town of Zhejiang Province

6.4.7 External links

- Environmental Activists Detained in Hangzhou
- Update about detention
- "Trial of Chinese environmental activist to start Monday"
- "China begins trial of environmentalist"

6.5 Wu Lihong

This is a Chinese name; the family name is 吴 (Wu).

Wu Lihong (simplified Chinese: 吴立红; traditional Chinese: 吳立紅; pinyin: *Wú Lìhóng*; born 1968) is an environmental activist of the People's Republic of China. In August 2007, Wu was sentenced to prison by a local court in

retribution for a 10-year crusade against pollution in Lake Tai. Ironically, the lake has been suffering from a "pond scum" outbreak since May, verifying Wu's claims that the government and big business were polluting and endangering the ecology of a water system that provides water for over 2 million people.

The New York Times ran an online article on his plight on 14 October 2007. An excerpt follows:

> "Mr. Wu, a jaunty, 40-year-old former factory salesman, pioneered a style of intrepid, media-savvy environmental work that made Lake Tai, and the hundreds of chemical factories on its shores, the focus of intense regulatory scrutiny.
>
> In 2005 he was declared an "Environmental Warrior" by the National People's Congress. His address book contained cellphone numbers for officials in Beijing and the provincial capital of Nanjing who outranked the party bosses where he lived.
>
> But Mr. Wu was far from untouchable. He lost his job. His wife lost hers. The police summoned, detained and interrogated him. The local government and factory owners also tried for years to bring him into the fold with contracts, gifts and jobs. When party officials offered him a chance to profit handsomely from a pollution cleanup contract, a friend warned him not to accept. Mr. Wu, who needed the money, said yes.
>
> The country's third largest freshwater body, Lake Tai, or Taihu in Chinese, has long provided the people of the lower Yangtze River Delta with both their wealth and their conception of natural beauty.
>
> It nurtured a bounty of the "three whites," white shrimp, whitebait and whitefish, and a freshwater crustacean delicacy called the hairy crab. Natural and man-made streams irrigated rice paddies, and a network of canals ferried that produce far and wide.
>
> Along the lake's northern reaches, near the city of Wuxi, placid waters and misty hills captured the imagination of Chinese for hundreds of years. The wealthy built gardens that featured the lake's wrinkled, water-scarred limestone rocks set in groves of bamboo and chrysanthemum."

6.5.1 Pollution of Lake Tai

In May 2007, the lake was overtaken by a major algal bloom. Authorities are blaming this on the lowest water levels in 50 years. However, low water levels alone did not cause these blooms. Increases in nutrients, from fertilizer for example, create conditions conducive to algae blooms, which has polluted the water with a toxic blue substance and a foul smell, making the water unusable. The Chinese government has called the lake a major natural disaster despite the clearly anthropogenic origin of this environmental catastrophe. With the average price of bottled water rising to six times the normal rate, the government has banned all regional water providers from implementing price hikes. [2] Wuxi, which draws its tap water from the lake, has been particularly badly affected. As of October 2007, the Chinese government had shut down or given notice to over 1,300 factories around the lake. Some say that only unprofitable factories have been closed, others view the anti-pollution move as overkill. Despite of Chinese authorities' increasing awareness of environmental problems, Wu Lihong was arrested and tried for alleged extortion of one of the polluters (see Economist article). He received a three-year prison sentence. Released on April 12, 2010, Wu Lihong told journalists that he suffered brutal treatment during his three-year detention.*[1]*[2]

6.5.2 See also

- Environment of China

- Tan Kai, a computer technician and an environmental activist from Zhejiang

6.5.3 References

[1] <http://news.yahoo.com/s/afp/20100511/wl_asia_afp/chinaenvironmentpollutionrights>.

[2] <http://www.lemonde.fr/international/article/2010/05/11/l-ecologiste-chinois-wu-lihong-raconte-ses-conditions-de-detention_1349367_3210.html>.

6.5.4 External links

- Joseph Kahn, "In China, a Lake's Champion Imperils Himself", *International Herald Tribune*, October 13, 2007

Chapter 7

Text and image sources, contributors, and licenses

7.1 Text

- **Environmental issues in China** *Source:* https://en.wikipedia.org/wiki/Environmental_issues_in_China?oldid=710989954 *Contributors:* Fred Bauder, William M. Connolley, Ijon, Alan Liefting, Shiftchange, Vsmith, Gary, Velella, Gruguuru, Rjwilmsi, Moe Epsilon, Emijrp, Arthur Rubin, タチコマ robot, Chris the speller, Sct72, Derek R Bullamore, Bellerophon5685, Heroeswithmetaphors, DASonnenfeld, Metal.lunchbox, Yintan, JL-Bot, Watti Renew, TheOldJacobite, Niceguyedc, Jeremiestrother, Light show, Jarble, Yobot, AnomieBOT, ThaddeusB, Anna Frodesiak, Hongsy, JayJay, Kaiyr, Diblidabliduu, RjwilmsiBot, EmausBot, Homunculus, Josve05a, AndrewOne, Donner60, ClueBot NG, CopperSquare, Newyorkadam, Rund Van, MusikAnimal, Mark Arsten, Darwinian Ape, Acadēmica Orientālis, Soulparadox, Solar Kermit, Mogism, Vanamonde93, EvergreenFir, ArchPope Sextus VI, DDDIIZZZYY, Monkbot, Rubbish computer, Elizben, Anticla rutila and Anonymous: 64

- **Environment of China** *Source:* https://en.wikipedia.org/wiki/Environment_of_China?oldid=707871347 *Contributors:* Olivier, Fred Bauder, Voidvector, Mac, Jiang, Fuzheado, RedWolf, Rursus, SchmuckyTheCat, Ghormax, Timrollpickering, Alan Liefting, Akadruid, Netoholic, Jackol, Edcolins, Formeruser-81, Antandrus, Beland, Oneiros, Huaiwei, Mike Rosoft, Discospinster, Rich Farmbrough, ESkog, Yongliu, Bobo192, Nsaa, Gary, Arthena, Kusma, Instantnood, WadeSimMiser, Wikiklrsc, Jasonkibby, MC MasterChef, Kbdank71, Mendaliv, Rjwilmsi, DoubleBlue, Nsae Comp, Gurch, Codex Sinaiticus, Benlisquare, Gwernol, Vmenkov, Wavelength, Ryz05, Gaius Cornelius, Wimt, Daveswagon, Badagnani, Rjensen, Bobak, Emesik, Nlu, Closedmouth, Sardanaphalus, SmackBot, Hu Gadarn, KVDP, Jab843, PJM, Brianski, Ohnoitsjamie, Hmains, Skizzik, Persian Poet Gal, Darth Panda, Krich, Cybercobra, Bowlhover, Dreadstar, Richard001, Swerz, Zero Gravity, Aaker, Kukini, Ohconfucius, Mukadderat, Anlace, Gobonobo, Jaganath, Mbeychok, JHunterJ, Noah Salzman, Martinp23, Hu12, 1989flame, Joseph Solis in Australia, Tawkerbot2, Le savoir et le savoir-faire, JohnCD, Keenan the sperry, Argon233, Old Guard, Neelix, Cydebot, Sinolonghai, Dancter, Kozuch, Joowwww, Mattisse, Epbr123, Mojo Hand, Ideogram, Holderlin, Heroeswithmetaphors, Niohe, Maork, Hermant patel, MER-C, Instinct, RiseOfTheRev01ution, Xeno, Snowolfd4, Magioladitis, Connormah, VoABot II, Mredheffer, Twsx, Engineman, Catgut, Animum, Cgingold, Allstarecho, Schumi555, Utc-100, Macop1, Paul Gard, MartinBot, Jonathan Hall, J.delanoy, Jorgenumata, Svetovid, Xue hanyu, Chrisvnicholson, Epborn, NewEnglandYankee, Burzmali, Idioma-bot, Agamemnus, VolkovBot, Amaraiel, Metal.lunchbox, Johnfos, Lop.dong, Ericdn, Indubitably, BillSung, PNG crusade bot, Seraphim, LeaveSleaves, Cewiz, Hofaizi, Zhenqinli, Choson4eva, Newsaholic, HybridBoy, Teh nubkilr, Moonriddengirl, WereSpielChequers, Jauerback, Keilana, Arbor to SJ, Oxymoron83, Lightmouse, Spitfire19, Gomeying, WikiLaurent, Li88wu8, JL-Bot, Esirec, MarsmanRom, ClueBot, Binksternet, The Thing That Should Not Be, Mookie25, Der Golem, Blanchardb, CohesionBot, Moreau1, Gary Yam, Tyler, NuclearWarfare, Techfast50, Hadooooookin, Redthoreau, Zappa711, God iz Ded, DumZiBoT, Humortueio, Chungster21, SilvonenBot, Knautzen~enwiki, Airplaneman, CalumH93, Vegetarianrage, Addbot, Ronhjones, Krypt0fish, Cut1664, CanadianLinuxUser, Glane23, Ld100, Pyl, Tide rolls, Lightbot, Jarble, Sta0018, Yobot, Xu Davella, THEN WHO WAS PHONE?, Pganas, Ayrton Prost, Burningjoker, Juzhong, Kristen Eriksen, ThaddeusB, Qajar, Arilang1234, Ermingtonboi, Materialscientist, Gilbert04, High Contrast, Neurolysis, Natandoron9, Ikaria-Tennison, Volvo B9TL, Chongkian, Howsa12, Sarwicked, Sector001, FrescoBot, Aryoerg, Trust Is All You Need, I dream of horses, Elockid, Serols, Zhonghuo~enwiki, Sfilmsactiwo, Benefactordyr, Tvenell, Jarmihi, RjwilmsiBot, Bento00, VernoWhitney, Sophie2895, Polylepsis, Acather96, YellowPops, Ascheineson, Rédacteur Tibet2, 23 shanghai metro, China Dialogue Net, L Kensington, Inadr3am, Imperratoor, Ik9939, ClueBot NG, Pizzaman13579, Sir Moose, Jokeroler, Snotbot, Onghuiqi89, U0802701, Cameronbuettner, Electriccatfish2, Cold Season, Gallagher-photo, RudolfRed, Lukas[23], Acadēmica Orientālis, Soulparadox, Joeinwiki, Secondhand Work, Lolol11111 and Anonymous: 263

- **Debate over China's economic responsibilities for climate change mitigation** *Source:* https://en.wikipedia.org/wiki/Debate_over_China'{}s_economic_responsibilities_for_climate_change_mitigation?oldid=707779846 *Contributors:* Mac, Tpbradbury, SchmuckyTheCat, Timrollpickering, Alan Liefting, Chowbok, D6, Vsmith, Woohookitty, BD2412, Ground Zero, Benlisquare, Vmenkov, Wavelength, Arthur Rubin, NeilN, Sardanaphalus, SmackBot, PeterSymonds, Theanphibian, Gobonobo, Joseph Solis in Australia, CmdrObot, Cydebot, Sinolonghai, JohnInDC, Engineman, AstroHurricane001, Squids and Chips, Malik Shabazz, TheOldJacobite, Niceguyedc, Nymf, Nukeless, CathCarey, Kbdankbot, Yobot, Themfromspace, Lilpipsqueak628, Ikaria-Tennison, FrescoBot, Kiwikibble, Wgswanson, DrilBot, Hai398, 775852O, Huaxia, FelixtheMagnificent, RjwilmsiBot, SporkBot, Helpful Pixie Bot, BattyBot, Acadēmica Orientālis, Cyberbot II, RotlinkBot, Monkbot, Stewi101015 and Anonymous: 82

- **Desertification in China** *Source:* https://en.wikipedia.org/wiki/Environmental_issues_in_China?oldid=710989954 *Contributors:* Fred Bauder, William M. Connolley, Ijon, Alan Liefting, Shiftchange, Vsmith, Gary, Velella, Gruguuru, Rjwilmsi, Moe Epsilon, Emijrp, Arthur Rubin, タ チコマ robot, Chris the speller, Sct72, Derek R Bullamore, Bellerophon5685, Heroeswithmetaphors, DASonnenfeld, Metal.lunchbox, Yintan, JL-Bot, Watti Renew, TheOldJacobite, Niceguyedc, Jeremiestrother, Light show, Jarble, Yobot, AnomieBOT, ThaddeusB, Anna Frodesiak, Hongsy, JayJay, Kaiyr, Diblidabliduu, RjwilmsiBot, EmausBot, Homunculus, Josve05a, AndrewOne, Donner60, ClueBot NG, CopperSquare, Newyorkadam, Rund Van, MusikAnimal, Mark Arsten, Darwinian Ape, Acadēmica Orientālis, Soulparadox, Solar Kermit, Mogism, Vanamonde93, EvergreenFir, ArchPope Sextus VI, DDDIIZZZYY, Monkbot, Rubbish computer, Elizben, Anticla rutila and Anonymous: 64

- **Dongtan** *Source:* https://en.wikipedia.org/wiki/Dongtan?oldid=704277935 *Contributors:* Genie, Bearcat, Tlogmer, Wertperch, Kmsiever, Babelfisch, Smyth, Howrealisreal, Fleetham, SMC, Faduci, MiracleMat, Gurch, Wavelength, CaliforniaAliBaba, Varano, Caerwine, Hmains, Bluebot, Rhollenton, Frap, Gobonobo, Cajolingwilhelm, Cydebot, Alaibot, Joowwww, Gralo, Heroeswithmetaphors, Fayenatic london, Gzli888, The Anomebot2, Escier~enwiki, Mcnattyp, MartinBot, Kevinsam, Xue hanyu, Tuduser, Gruschke, Andy Johnston, Davecrosby uk, GrahamHardy, Johnfos, Falcon8765, Regregex, LeadSongDog, Altzinn, TFCforever, DMcKendrick, Cambrasa, Auntof6, Excirial, PixelBot, Zappa711, Addbot, Pauldingxi, Sh1019, LlywelynII, Materialscientist, AddisWang, Bethhurran, Payattention007, FrescoBot, Jonesey95, Jaguar, Serols, Hai398, ZhBot, Reach Out to the Truth, RjwilmsiBot, Truehubster, Philippk, せいきょういく, Puffin, Petrb, ClueBot NG, Widr, The Banner Turbo, Sawol, Acadēmica Orientālis, ChrisGualtieri, Graphium, Lemnaminor, GingerGeek, ArchPope Sextus VI, Dongtan7557, SPIKE SPIKE BAD, Ya dad 123, CasparLee123 and Anonymous: 87

- **Huangbaiyu** *Source:* https://en.wikipedia.org/wiki/Huangbaiyu?oldid=704281448 *Contributors:* Alan Liefting, Mrtrey99, Alai, Rjwilmsi, Agamemnon2, Atwood, SmackBot, Hmains, Nick carson, Felipec, Mr3641, Heroeswithmetaphors, Jllm06, The Anomebot2, Kevinsam, Hoyohoyogold, LethalTuna, MystBot, Addbot, Kimmelm, Green Cardamom, FrescoBot, Jaguar, ZhBot, H3llBot, Erianna, MoondyneAWB, Acadēmica Orientālis, ChrisGualtieri, Monkbot and Anonymous: 8

- **Qidong protest** *Source:* https://en.wikipedia.org/wiki/Qidong_protest?oldid=694310477 *Contributors:* Davidcannon, Alan Liefting, Cattus, Ohconfucius, Cydebot, Aiko, Addbot, FreeRangeFrog, Homunculus, Charles Essie, Eakopskvm, Lostromantic and Haonan.zhou

- **Shifang protest** *Source:* https://en.wikipedia.org/wiki/Shifang_protest?oldid=638925539 *Contributors:* Pigsonthewing, Wavelength, Chris the speller, Cattus, Novangelis, Cydebot, EmausBot, Homunculus, Jarodalien, Quick and Dirty User Account, Begonia Brandbygeana, Khazar2, Dexbot, Charles Essie, 漢浩, Jonlau22800 and Anonymous: 2

- **Three-North Shelter Forest Program** *Source:* https://en.wikipedia.org/wiki/Three-North_Shelter_Forest_Program?oldid=709183138 *Contributors:* The Anome, Menchi, Selket, Shizhao, Quadalpha, Alan Liefting, Rich Farmbrough, Markussep, Bobo192, Anthony Appleyard, Alai, BD2412, Chobot, Vmenkov, Eraserhead1, Pigman, Dialectric, Robyvecchio, Arthur Rubin, SmackBot, Hmains, CSWarren, Jeff5102, Rigadoun, 041744, Joseph Solis in Australia, Teratornis, JuWiki, Stefan Jensen, Ilovetuna, The Anomebot2, Engineman, Cgingold, Chauncey freak, Balthazarduju, DASonnenfeld, Hugo999, TXiKiBoT, Karmos, Piperh, Naive rm, Mardhil, Crash Underride, JuWiki2, Lightmouse, ClueBot, Coinmanj, 7, Dana boomer, Tdslk, WikHead, Airplaneman, Addbot, Lightbot, Wikimono111, Poko, Apothecia, Cureden, Ita140188, Thehelpfulbot, LucienBOT, Pinethicket, Elekhh, ZhBot, RjwilmsiBot, EmausBot, ECTaiwan2010, ZéroBot, Bxj, MRant71295, ClueBot NG, Michaelmas1957, Gob Lofa, Cold Season, Vandhana297, BattyBot, Acadēmica Orientālis, ChrisGualtieri, Monkbot, Cailynjhkim and Anonymous: 58

- **Water resources of China** *Source:* https://en.wikipedia.org/wiki/Water_resources_of_China?oldid=711297470 *Contributors:* Fred Bauder, Timrollpickering, Alan Liefting, Confuzion, Yuje, Gene Nygaard, Instantnood, Macaddct1984, Rjwilmsi, Ligulem, Vmenkov, Kimchi.sg, Epipelagic, Sardanaphalus, SmackBot, Hmains, Skizzik, Cybercobra, Neelix, Cydebot, Ideogram, Niohe, AntiVandalBot, MER-C, Mschiffler, CommonsDelinker, Erkan Yilmaz, Xue hanyu, Hugo999, Lop.dong, Vitund, Yintan, Aspects, ClueBot, Der Golem, Watti Renew, Doseiai2, Chin.mu, Htes.nehoc, Gary Yam, Hadoooookin, Thewellman, Cut1664, Jasper Deng, Jarble, EdwardLane, Materialscientist, Mengmeng12388, Armbrust, Full-date unlinking bot, Zhonghuo~enwiki, John of Reading, Super48paul, 7partparadigm, Batemqi4382, ClueBot NG, Sawol, Acadēmica Orientālis, Arcandam, Nona noor and Anonymous: 25

- **China Beijing Environmental Exchange** *Source:* https://en.wikipedia.org/wiki/China_Beijing_Environmental_Exchange?oldid=682670329 *Contributors:* Tabletop, Arthur Rubin, Cattus, Cydebot, Magioladitis, DASonnenfeld, Jx-10, Acadēmica Orientālis and Anonymous: 1

- **Climate change in China** *Source:* https://en.wikipedia.org/wiki/Climate_change_in_China?oldid=705959809 *Contributors:* Mac, Alan Liefting, Vsmith, Ground Zero, Eraserhead1, Arthur Rubin, Ohconfucius, Dl2000, R'n'B, AstroHurricane001, DASonnenfeld, Nopetro, Watti Renew, TheOldJacobite, Lightbot, Jarble, Cmano13, Yobot, AnomieBOT, LilHelpa, Anna Frodesiak, Erik9bot, Kwiki, RockfangSemi, I dream of horses, Hai398, GreenpeaceChina~enwiki, RjwilmsiBot, John of Reading, Dewritech, Hamiltha, ClueBot NG, Widr, Antiqueight, BG19bot, Northamerica1000, Acadēmica Orientālis, EagerToddler39, Wai0004, Joeinwiki, Epicgenius, Jamesmcmahon0, AmaryllisGardener, Climate123, Genesis0526, ShiLanruo, Cailynjhkim, Climate supporter, Tallymarker and Anonymous: 51

- **Coal in China** *Source:* https://en.wikipedia.org/wiki/Coal_in_China?oldid=714576152 *Contributors:* Maury Markowitz, Fred Bauder, Mac, Julesd, Timrollpickering, Aetheling, Alan Liefting, Wwoods, PFHLai, Icairns, Rich Farmbrough, Vsmith, Eric Kvaalen, Vsion, Wavelength, NawlinWiki, TheSeer, Arthur Rubin, Katieh5584, Shubi, Sardanaphalus, SmackBot, Hmains, Ottawakismet, Cattus, Hibernian, BrianBird, Theanphibian, Peterlewis, Hvn0413, Petr Matas, Keenan the sperry, Neelix, Cydebot, Sinolonghai, Marek69, Aiko, Peace01234, Danger, OSX, Hermant patel, Montie5, Beagel, JaGa, E-pen, CommonsDelinker, Parkerconrad, SriMesh, KudzuVine, Squids and Chips, Xenonice, Oshwah, Autozavod, Natandoron, Plazak, Staka, Swliv, Unregistered.coward, Nopetro, Lightmouse, ClueBot, Watti Renew, Arbeit Sockenpuppe, Arjayay, Techfast50, Nukeless, DumZiBoT, Temeku, Quetssef, Scully42, Addbot, Dave606, FR45804, Luckas-bot, Yobot, LLTimes, AnomieBOT, Natandoron9, Jsharpminor, Ikaria-Tennison, Ita140188, Frozenevolution, Trust Is All You Need, DivineAlpha, Meduenin, Full-date unlinking bot, Sfilmsactiwo, Orangesoda123, Fiftytwo thirty, In ictu oculi, TGCP, EmausBot, Shanghai 10, Dewritech, ZéroBot, Jenks24, H3llBot, LordkanzlerAN, Шиманський Василь, ClueBot NG, ProgressiveThinker, Ksid-3k, Frietjes, Geistcj, Monicatan GPEA, RudolfRed, BattyBot, Cyberbot II, Mogism, Cerabot~enwiki, Nadya Inoubli, Paul Bodnick, Mak910, Monkbot, Vieque, ShiLanruo, Skoritz, Wzr123, Muzzleflash, Ysalamari, BoyBoom and Anonymous: 59

- **Tianjin Climate Exchange** *Source:* https://en.wikipedia.org/wiki/Tianjin_Climate_Exchange?oldid=546328115 *Contributors:* Arthur Rubin, Montie5, R'n'B, TXiKiBoT, Trust Is All You Need, Hajatvrc, EmausBot, Liangfangyu and Anonymous: 3

- **Center for Legal Assistance to Pollution Victims** *Source:* https://en.wikipedia.org/wiki/Center_for_Legal_Assistance_to_Pollution_Victims?oldid=648170138 *Contributors:* Kleinzach, Hu12, Magioladitis, Cgingold, DASonnenfeld, Alvin Seville, Howchou, GoingBatty, Tanzeelahad, PRC.USA and Anonymous: 1

- **China Pollution Map Database** *Source:* https://en.wikipedia.org/wiki/China_Pollution_Map_Database?oldid=679699170 *Contributors:* Alan Liefting, Wavelength, Wikimedes, SwisterTwister, AnomieBOT, I dream of horses, ClueBot NG, TaoliangIPE, Galant Khan and Anonymous: 3

- **JSYU UAV** *Source:* https://en.wikipedia.org/wiki/JSYU_UAV?oldid=647207545 *Contributors:* Wavelength, XdeLaTorre, Jj98, BattyBot and Anonymous: 1

- **Pollution in China** *Source:* https://en.wikipedia.org/wiki/Pollution_in_China?oldid=714479925 *Contributors:* Ubiquity, Fred Bauder, Fuzheado, Hawstom, Alan Liefting, Alexf, SimonLyall, Discospinster, Smyth, Smalljim, Ogress, Alansohn, Gary, Mduvekot, BDD, GregorB, JHMM13, DVdm, Bgwhite, Wavelength, Rsrikanth05, Antsun85, Aaron Brenneman, Daniel Mietchen, Bdell555, Arthur Rubin, SmackBot, McGeddon, Gilliam, Chris the speller, Egsan Bacon, Bejnar, Ohconfucius, Christian75, Jamesjiao, Spencer, Uchohan, Superjoo, Cgingold, CommonsDelinker, NewEnglandYankee, DASonnenfeld, Metal.lunchbox, Fences and windows, Pjrobertson, Oshwah, Meters, SieBot, Mikemoral, Smsarmad, Flyer22 Reborn, Denisarona, Mr. Granger, ClueBot, Excirial, Muhandes, M.O.X, Lily1104, Lightshow, PCHS-NJROTC, Jax 0677, XLinkBot, Karpouzi, Addbot, Ethanpet113, Lihaas, Jarble, Yobot, AzureFury, WatcherZero, AnomieBOT, Stormedelf, Materialscientist, Xqbot, Anna Frodesiak, Amaury, Edward-F, RightCowLeftCoast, George2001hi, FrescoBot, LucienBOT, Pinethicket, I dream of horses, Serols, Kaiyr, Phoenix7777, Full-date unlinking bot, Double sharp, Zanhe, Dxhunzai, Onel5969, RjwilmsiBot, John of Reading, Immunize, Dewritech, RA0808, Dcirovic, K6ka, Cymru.lass, Makecat, Rcsprinter123, Labnoor, Donner60, Petrb, ClueBot NG, Satellizer, Widr, Ryan Vesey, Helpful Pixie Bot, MusikAnimal, Mifter Public, Amp71, FutureTrillionaire, NLG900, Lzy881114, BattyBot, David.moreno72, Acadēmica Orientālis, Cyberbot II, SmileyLlama, Dexbot, Lugia2453, Thisisahacker, Exenola, 63grak63, DavidLeighEllis, Killerboy30, Ugog Nizdast, E8xE8, Manul, Ffffafsd, Adachoq, Icensnow42, LegendRemastered, JaconaFrere, Violincrazydude13, CatcherStorm, Lagoset, Davidwils, Chorrall, EnglishdDJ, Minor Syntax Edits, Misaugstad, CALGARY ROCKS, ShiLanruo, M semione, Vidalius, CAPTAIN RAJU, AWildAppeared, Daveeeeee4, Brendansinnamon and Anonymous: 162

- **SDAS UAV** *Source:* https://en.wikipedia.org/wiki/SDAS_UAV?oldid=714235091 *Contributors:* Bgwhite, Wavelength, Chris the speller, XdeLaTorre, BattyBot and Anonymous: 2

- **Yunnan Jinding Zinc** *Source:* https://en.wikipedia.org/wiki/Yunnan_Jinding_Zinc?oldid=695190117 *Contributors:* Fred Bauder and Davidcannon

- **EcoPark (Hong Kong)** *Source:* https://en.wikipedia.org/wiki/EcoPark_(Hong_Kong)?oldid=440818913 *Contributors:* Bobblewik, Andycjp, Vortexrealm, Instantnood, HenryLi, K3ith, Carbonix, Huibe~enwiki, Oneironautdawn, Tawkerbot2, Mr3641, Insanephantom, Peter Chastain, Lamro, Mothers.dreams and Anonymous: 4

- **Electronic waste in China** *Source:* https://en.wikipedia.org/wiki/Electronic_waste_in_China?oldid=660934717 *Contributors:* Alan Liefting, Chris Capoccia, Brz7, TenPoundHammer, ThomasO1989, Magioladitis, Wiae, Yobot, AnomieBOT, Citation bot, LilHelpa, I dream of horses, GoingBatty, ClueBot NG, Widr, Acadēmica Orientālis, Lillian08, Wikimarnie, Monkbot, Julien.bettler, Butric and Anonymous: 12

- **Electronic waste in Guiyu** *Source:* https://en.wikipedia.org/wiki/Electronic_waste_in_Guiyu?oldid=715087863 *Contributors:* Kierant, Jni, Auric, Alan Liefting, Tom, Leondz, Mendaliv, Wavelength, Tommyt, Brz7, SmackBot, Ctashian, Centrepull, Jimjamjak, CommonsDelinker, Leyo, TXiKiBoT, Falcon8765, Michael Frind, Flyer22 Reborn, Crywalt, John Nevard, Lily1104, Koumz, Rror, Addbot, Myk60640, Teles, Jarble, The Bushranger, Yobot, AnomieBOT, Citation bot, Andrewmin, J04n, Amaury, Safiel, Z50rbandit, Haeinous, Elspru, Ekpearce, Wchsunc1, Ὀ οἶστρος, Sonicyouth86, ClueBot NG, Irrigator, BG19bot, BattyBot, Solntsa90, Acadēmica Orientālis, CrunchySkies, Kwisha, Ahhblay, Harlem Baker Hughes, Stormmeteo, Adriano G. V. Esposito, 3298230932782302, OSAMASINLADLE, FilthyFranku and Anonymous: 33

- **Food and Environmental Hygiene Department** *Source:* https://en.wikipedia.org/wiki/Food_and_Environmental_Hygiene_Department?oldid=704017914 *Contributors:* Bourquie, Vortexrealm, Minghong, Fat pig73, Instantnood, HenryLi, Rjwilmsi, Winhunter, Benjwong, Sardanaphalus, SmackBot, Takamaxa, Nxn 0405 chl, Grumpyyoungman01, Shrimp wong, Ericlaw02, Raphaelhui, MainlyTwelve, Connormah, Philg88, Wylve, Rayyung, Addbot, Misterx2000, Tavatar, Chongkian, FrescoBot, ZhBot, Citobun and Anonymous: 5

- **Gin Drinkers Bay** *Source:* https://en.wikipedia.org/wiki/Gin_Drinkers_Bay?oldid=706099556 *Contributors:* Olivier, Jogloran, Vortexrealm, HenryLi, Bluemoose, Winhunter, KX675, Wavelength, SmackBot, Ricky@36, Takamaxa, Park3r, Mike2525, HongQiGong, Sam Li, CmdrObot, Hylas Chung, The Anomebot2, Philg88, Kevinsam, Jerry, Martarius, Addbot, Underwaterbuffalo, Lightbot, Yobot, ZhBot, Citobun and Anonymous: 2

- **Inventory of hazardous materials** *Source:* https://en.wikipedia.org/wiki/Inventory_of_hazardous_materials?oldid=704281864 *Contributors:* Bearcat, Wavelength, Ohconfucius, ShelfSkewed, R'n'B, Katharineamy, Hugo999, Mild Bill Hiccup, Yobot, AnomieBOT, Tavatar, Alvin Seville, Jaguar, Ratava smot, Armstrong10, Calvalshrk and Anonymous: 2

- **Kwai Chung Incineration Plant** *Source:* https://en.wikipedia.org/wiki/Kwai_Chung_Incineration_Plant?oldid=706100434 *Contributors:* Olivier, Alan Liefting, Vortexrealm, Minghong, Instantnood, HenryLi, Vegaswikian, Bgwhite, Foxxygirltamara, SmackBot, Hydrogen Iodide, Chris the speller, Claush66, Vina-iwbot~enwiki, Tripbeetle, The Anomebot2, Beagel, Philg88, Kevinsam, Louisamild, ClueBot, George955, Pakaraki, Addbot, Underwaterbuffalo, Lightbot, Vzty, ChuispastonBot, BG19bot, ArmbrustBot and Anonymous: 7

- **Regional Council (Hong Kong)** *Source:* https://en.wikipedia.org/wiki/Regional_Council_(Hong_Kong)?oldid=697693714 *Contributors:* Chrism, Mintchocicecream, Davidcannon, Alan Liefting, Vortexrealm, Giraffedata, Deryck Chan, Mcy jerry, Instantnood, HenryLi, Winhunter, YurikBot, Hinto, Cetot, OhanaUnited, Wylve, AlleborgoBot, Lmmnhn, Addbot, Douglas the Comeback Kid, Tavatar, Date delinker, ZhBot, Dewritech, Frietjes, Citobun and Anonymous: 6

- **Sai Tso Wan Recreation Ground** *Source:* https://en.wikipedia.org/wiki/Sai_Tso_Wan_Recreation_Ground?oldid=692834613 *Contributors:* MisfitToys, Deryck Chan, Josh Parris, Npeters22, SmackBot, ERcheck, JohnCD, Bobblehead, Miller17CU94, The Anomebot2, Ngchikit, KylieTastic, Deor, Lamro, Addbot, Underwaterbuffalo, KamikazeBot, Kam47625, John of Reading and Anonymous: 2

- **Sha Tin Sewage Treatment Works** *Source:* https://en.wikipedia.org/wiki/Sha_Tin_Sewage_Treatment_Works?oldid=607345660 *Contributors:* Rjwilmsi, Insanephantom, The Anomebot2, Philg88, Kevinsam, Underwaterbuffalo, Yobot, EmausBot, Cyberbot II and Anonymous: 1

- **Urban Council** *Source:* https://en.wikipedia.org/wiki/Urban_Council?oldid=709530029 *Contributors:* Mintchocicecream, SchmuckyTheCat, Ghormax, Vortexrealm, Deryck Chan, Mcy jerry, Alai, Instantnood, HenryLi, Alanmak, Winhunter, Ysw1987, YurikBot, Hinto, Cetot, Commander Keane bot, Gilliam, Bluebot, OrphanBot, Ohconfucius, Joseph Solis in Australia, HongQiGong, Jed, Tokyogirl79, Clithering, Fionli,

AlleborgoBot, Flyer22 Reborn, Lmmnhn, Niceguyedc, Rror, Addbot, Douglas the Comeback Kid, West.andrew.g, Lightbot, AnomieBOT, Xqbot, Date delinker, ZhBot, Ripchip Bot, EmausBot, Dewritech, Onanoff, MelbourneStar, Frietjes, Ho Yi, Lesser Cartographies, Citobun and Anonymous: 14

- **Waste management in Hong Kong** *Source:* https://en.wikipedia.org/wiki/Waste_management_in_Hong_Kong?oldid=705049343 *Contributors:* Davidcannon, Alan Liefting, Rich Farmbrough, Cmdrjameson, Rjwilmsi, Bgwhite, SmackBot, Gilliam, Richard001, A. Parrot, BranStark, Superzohar, JamesBWatson, Twsx, Wylve, R'n'B, Jamesontai, DASonnenfeld, Squids and Chips, ClueBot, The Thing That Should Not Be, Der Golem, Techfast50, DumZiBoT, Lilinaran, Underwaterbuffalo, Misterx2000, Douglas the Comeback Kid, VigilanteVern, Yobot, AnomieBOT, Materialscientist, LilHelpa, RenegadeMonster, Citation bot 1, DrilBot, Some Color Mage, BattyBot, RotlinkBot, Mseunicewong, Davidwils, BethNaught, AusLondonder and Anonymous: 18

- **Animal welfare and rights in China** *Source:* https://en.wikipedia.org/wiki/Animal_welfare_and_rights_in_China?oldid=714381254 *Contributors:* Rathfelder, Chris the speller, JzG, Keith D, Sega31098, AnomieBOT, LilHelpa, Rasnaboy, DexDor, Turn685, NewWorld101, DrChrissy, Brian Tomasik, BattyBot, The Original Filfi, Coejnk, Anticla rutila and Anonymous: 6

- **Environmental policy in China** *Source:* https://en.wikipedia.org/wiki/Environmental_policy_in_China?oldid=708425314 *Contributors:* Kku, Tpbradbury, Wetman, Davidcannon, Vsmith, A D Monroe III, Badagnani, DASonnenfeld, Liuzhou, Der Golem, Jarble, AnomieBOT, ThaddeusB, Xqbot, Trappist the monk, Wikipelli, AvicBot, ClueBot NG, Cyberbot II, PraetorianFury, B14709, 彭家杰, 赵睿智, Monkbot, Jayakumar RG, Nonesquare, Elizben and Anonymous: 16

- **One-child policy** *Source:* https://en.wikipedia.org/wiki/One-child_policy?oldid=715875558 *Contributors:* AxelBoldt, Mav, Ed Poor, Roadrunner, Ark~enwiki, Formulax~enwiki, Retostamm, Hephaestos, Olivier, Stevertigo, Edward, Michael Hardy, Fred Bauder, Matthewmayer, Menchi, Ixfd64, Miciah, Bueller 007, BigFatBuddha, Julesd, Jiang, Mxn, Technopilgrim, Charles Matthews, Lfh, N-true, Tedius Zanarukando, Colipon, Savantpol, Andrewman327, WhisperToMe, Saltine, Itai, Pakaran, Mina86, Shantavira, Ke4roh, Wanwan, ZimZalaBim, Peak, Naddy, Lowellian, SchmuckyTheCat, Sunray, Bkell, Moink, Hadal, JackofOz, Alan Liefting, Smjg, Nikodemos, Gil Dawson, Jpta~enwiki, Fastfission, Obli, Bradeos Graphon, No Guru, Brona, Andris, Cloud200, Gzornenplatz, Matthäus Wander, Slurslee, SoWhy, R. fiend, Formeruser-81, Antandrus, Beland, MarkSweep, Kiteinthewind, Xtreambar, Kesac, Gscshoyru, Huaiwei, Joyous!, MementoVivere, Flex, Mike Rosoft, Jayjg, Samf-nz, Freakofnurture, Miborovsky, Arcataroger, Discospinster, Rich Farmbrough, R6144, Smyth, User2004, David Schaich, Tsujigiri~enwiki, Bender235, Rubicon, ESkog, Kbh3rd, Mcpusc, Violetriga, Ylee, Purplefeltangel, El C, Shrike, Triona, Aaronbrick, Coolcaesar, Bobo192, Stesmo, Smalljim, Viriditas, Adrian~enwiki, ParticleMan, Minghong, Nsaa, Mrzaius, Alansohn, Ctande, Arthena, Trainik, Cdc, Rwendland, NTK, Snowolf, Melaen, CaseInPoint, DanShearer, Evil Monkey, Toytown Mafia, W7KyzmJt, Instantnood, LukeSurl, HenryLi, Dejvid, Kelly Martin, Starblind, Mel Etitis, Mindmatrix, ScottDavis, LOL, Daniel Case, Davidkazuhiro, WadeSimMiser, Duncan.france, Apokrif, USSJoin, Dolfrog, Paradon, Prashanthns, Graham87, BD2412, MC MasterChef, RadioActive~enwiki, Search4Lancer, Sjö, Rjwilmsi, Nightscream, SMC, Funnyhat, Jdowland, Boccobrock, Durin, Bhadani, Ttwaring, Plastictv, Yamamoto Ichiro, FlaBot, Skyfiler, Latka, Winhunter, Nihiltres, Vsion, Nivix, Payo, Gurch, Nimur, Pevernagie, Alphachimp, Sdr, Gurubrahma, Cpcheung, Benjwong, Mark Yen, Theshibboleth, Benlisquare, DVdm, Volunteer Marek, Tonync, Bgwhite, Kakurady, FrankTobia, Banaticus, Vmenkov, EamonnPKeane, The Rambling Man, Wavelength, GrandCru, Eraserhead1, Ryz05, StuffOfInterest, Jeffhoy, John Smith's, Hornandsoccer, Severa, Taejo, Akamad, Stephenb, Shell Kinney, Gaius Cornelius, Ksyrie, CambridgeBayWeather, Pseudomonas, Wimt, Big Brother 1984, Daveswagon, GunnarRene, NawlinWiki, Majukutaj, Nutiketaiel, Daanschr, Dureo, Chooserr, Readparse, Anetode, RFBailey, Moe Epsilon, Theboogeyman, Zwobot, Syrthiss, Dbfirs, Mysid, TastyCakes, Jpeob, Gat0r, Wknight94, AjaxSmack, Zzuuzz, Wotnarg, Malekhanif, Theda, Closedmouth, E Wing, Josh3580, JoanneB, Vicarious, Barbutti, Cjwright79, Kevin, ArielGold, Dbarefoot, JeffBurdges, Allens, Katieh5584, Kungfuadam, LakeHMM, Lyrl, Kf4bdy, Tom Morris, That Guy, From That Show!, Dupz, TravisTX, AtomCrusher, Jsnx, SmackBot, 4dhayman, YellowMonkey, Haverpopper, Reedy, Shoy, Augest, Verne Equinox, Pennywisdom2099, Timeshifter, Canthusus, Kintetsubuffalo, Wzhao553, Typhoonchaser, Jack051, Siddyrocks, Yamaguchi 先生, Gilliam, Hmains, Skizzik, Armeria, Kazkaskazkasako, Chris the speller, Bluebot, Miquonranger03, Roscelese, SchfiftyThree, Tetraglot, TheLeopard, Quackslikeaduck, Colonies Chris, Darth Panda, Mexcellent, Remixed, Nick Levine, Shalom Yechiel, Cripipper, Nixeagle, Shibo77, Rrburke, Steelbeard1, Addshore, Midnightcomm, Jaimie Henry, CanDo, Cybercobra, Decltype, Dvc214, Dreadstar, Richard001, Derek R Bullamore, Hgilbert, A.V.~enwiki, Gbinal, DMacks, -Marcus-, Davidone, FelisLeo, Ohconfucius, Will Beback, Angela26, Krashlandon, Rklawton, Takamaxa, Sophia, KLLvr283, Soap, John, AmiDaniel, Jidanni, Heimstern, Edwy, Minna Sora no Shita, IronGargoyle, Ekrub-ntyh, Otav347, A. Parrot, Chuck Simmons, Agathoclea, Noah Salzman, Optakeover, Zabdiel, Fangfufu, Dammit, Dhp1080, Ambuj.Saxena, Condem, Robin Chen, NinjaCharlie, Zepheus, Mackan, BranStark, Iridescent, The Giant Puffin, Eric12, Joseph Solis in Australia, StephenBuxton, Fsotrain09, Ouzo~enwiki, HongQiGong, CapitalR, Amdurbin, Color probe, CharlesM, Courcelles, Audiosmurf, Tawkerbot2, Cryptic C62, Chen Zen, Lokiloki, JForget, Ale jrb, FunPika, Wafulz, Insanephantom, Zarex, Dycedarg, BeenAroundAWhile, Rawling, DavidCowhig, Benwildeboer, Djembe, Shultz IV, MrFish, MaxEnt, Funnyfarmofdoom, Yaris678, Cydebot, Briankim, Vanished user 2340rujowierfj08234irjwfw4, Sinolonghai, Mr.weedle, Gogo Dodo, Anthonyhcole, Barracuda57, Hanfresco, Shirulashem, HitroMilanese, Christian75, Codetiger, DumbBOT, The Lake Effect, Hontogaichiban, Omicronpersei8, Vanished User jdksfajlasd, Zalgo, Artur Buchhorn, Kirk Hilliard, BCSWowbagger, Ajsfiremouse, JamesAM, QueenE92, Thijs!bot, Epbr123, IvanStepaniuk, Qwyrxian, Dasani, Daniel, N5iln, Andyjsmith, Louis Waweru, James086, Bobby fletcher, Tellyaddict, Vchao, Sad mouse, Danielfolsom, PhiLiP, AntiVandalBot, Luna Santin, Seaphoto, Esprit Fugace, C.anguschandler, Pwhitwor, Mackan79, Maork, Mack2, Âme Errante, Dylan Lake, Amberina, Gdo01, DarthShrine, Alphachimpbot, BenMcLean, Leuqarte, Res2216firestar, Leuko, AniRaptor2001, MER-C, The Transhumanist, FarnhamJ, Andonic, Doctorhawkes, Joshua, LittleOldMe, Magioladitis, Bakilas, Bongwarrior, VoABot II, Kuyabribri, Yandman, FrankSui, Yyyikes, Bcrawf, CTF83!, WODUP, Nick Cooper, Fleagle11, Catgut, Indon, Tasermon's Partner, Simonxag, 28421u2232nfenfcenc, Afaprof01, Mkdw, Schumi555, Cpl Syx, DerHexer, JaGa, Patstuart, RichMac, FisherQueen, MartinBot, AussieBoy, Ultraviolet scissor flame, Anaxial, Jay Litman, Xanon, AlexiusHoratius, Ash, Spaceetweek, PrestonH, Tgeairn, RockMFR, J.delanoy, Pharaoh of the Wizards, Truncated, Trusilver, Yonidebot, Jonpro, Neonuke, LedRush, Man Fung, Tommy11111, Mamyles, Darth Mike, Shawn in Montreal, Mikael Häggström, Gurchzilla, Jiu9, Blackraven42, M-le-mot-dit, Lyndond, NewEnglandYankee, BlGene, Jeff F F, Olegwiki, Hkb, KylieTastic, Cometstyles, Webslinger92, Jamesofur, DMCer, Lettik, WinterSpw, Jarry1250, Useight, Martial75, Trent31, Tyrant T100, Heias, 386-DX, Whatfg, VolkovBot, Science4sail, Indubitably, Kevinkor2, Philip Trueman, Frigglinn~enwiki, TXiKiBoT, Oshwah, Tavix, Vipinhari, GcSwRhIc, Someguy1221, Wesleykid, Tader1, Melsaran, Slysplace, Broadbot, LeaveSleaves, Amog, Mannafredo, Wiae, Ben Ward, Saturn star, Peter K Burian, Theli0nh3art, Falcon8765, Enviroboy, Thehararite, Why Not A Duck, Vscel4, Onceonthisisland, Adamoppe, Demize, Hokie92, LGNG, SylviaStanley, AmigoNico, SieBot, Tom t 117, Alessgrimal, Tiddly Tom, AS, Scarian, Euryalus, Laoris, Jauerback, Dawn Bard, Caltas, Mister1nothing, Raspberrysnapple, Triwbe, Eps200, Yintan, Emperor001, WRK, Bentogoa, Happysailor, Flyer22 Reborn, Bricktamland 05, DanEdmonds, Rédacteur Tibet, Oxymoron83, Avnjay, Harry-, Poindexter

Propellerhead, Boromir123, KathrynLybarger, Hobartimus, PalaceGuard008, Int21h, Einsteins37, Mitch1981, Johnanth, Asdfasdf1231234, StaticGull, Damientheomen3, Tesi1700, Mygerardromance, Jonfinkasu, Dust Filter, Pinkadelica, Escape Orbit, JoeenNc, ImageRemovalBot, ClueBot, Bwfrank, Strongsauce, Leahbookworm, Avenged Eightfold, J fong, The Thing That Should Not Be, B1atv, Voxpuppet, IceUnshattered, Vir Novus, Farolif, Franamax, Winger84, CounterVandalismBot, Niceguyedc, Otolemur crassicaudatus, Neverquick, Rogswikipage, Thegargoylevine, William Ortiz, Somno, Excirial, Jusdafax, Alibipie, Rorjam, Erebus Morgaine, Gtstricky, Vanhoabui, Benrcowan, Cenarium, Wilsmithy, BlueMooon~enwiki, Razorflame, Dekisugi, Gundersen53, Ottawa4ever, La Pianista, Thingg, Versus22, MelonBot, SoxBot III, Liberal Humanist, DumZiBoT, Georgexx316, Loranchet, Istib, XLinkBot, Fastily, SwirlBoy39, Honeybabs, Saltybaps, Jakobkennedy, Nepenthes, Little Mountain 5, Mitch Ames, Skarebo, CapnZapp, Changeidea, Badgernet, Jd027, Noctibus, Owlservice, ZooFari, Zodon, Dwilso, HexaChord, Thebestofall007, Fj217, Darkshot27, Fabio vaccaro123, Addbot, Rorybob, Roentgenium111, Njm0, Sketerpot, Some jerk on the Internet, DOI bot, Tryanmax, Non-dropframe, Captain-tucker, Duke of Chutney, I am not a dog, Binary TSO, Zzzaaaamm87, Ronhjones, MartinezMD, Jncraton, Moosehadley, CanadianLinuxUser, Zinger12357, TheGimpMan, Download, CarsracBot, DFS454, Glane23, Debresser, Charley-Contagious, DanaStreet, Blaylockjam10, Patton123, Dman1300, Hidude965, Newfraferz87, Jbennett8000, Tide rolls, BrianKnez, Lightbot, Krano, QuadrivialMind, Gail, Weaseloid, Jarble, Tallorno, Matt.T, Legobot, Kurtis, Luckas-bot, Yobot, AzureFury, Fraggle81, Ojay123, THEN WHO WAS PHONE?, Before I Die, AnakngAraw, Plasticbot, Eric-Wester, Tempodivalse, Synchronism, Bility, Backslash Forwardslash, AnomieBOT, Dirjarmocksorz, Whittlepedia, ThaddeusB, Rjanag, Killiondude, Jim1138, Barliman Butterbur, Fahadsadah, Kingpin13, Yachtsman1, ΘΕΟΔΩΡΟΣ, Flewis, Materialscientist, Citation bot, Bellemonde, Brightgalrs, Wranadu2, Frankenpuppy, Marcus014, Xqbot, Thankswiki55, Skjohnson1, Capricorn42, Bihco, Bianca1975, Etoombs, ANZAN, Chickamo, Arh3sfu, Kuros81, Gingermin002, Anonymous from the 21st century, GrouchoBot, Abce2, Shi Gelei, Annalise, Earlypsychosis, Wwbread, Editfinder, Sophus Bie, Shadowjams, პაპუა ქ, James1011R, Maxd1512, Shamengardner, Knightwhosayni, Tobby72, Sky Attacker, Lonaowna, Jc3s5h, VS6507, Bcefjj805, Jjanssen11, Troglo, Vishnu2011, Jamesooders, A little insignificant, HamburgerRadio, Citation bot 1, Cookie5500, Krisz987654321, DrilBot, Gautier lebon, Pinethicket, I dream of horses, Elockid, Arctic Night, 10metreh, Jonesey95, A412, Lazy Sk8, Jschnur, RedBot, LegendFPS, Serols, Σ, Dancopg1, Phoenix7777, Monkeymanman, Xiaoshan Math, Dude1818, Pristino, Reconsider the static, Colchester121891, Aarondimi, Yunshui, Dinher, Mono, Zanhe, Xinwen, Sky4mky, Lotje, GregKaye, Zebrabubbles, Dustynyfeathers, Vrenator, ChartaxS, Seahorseruler, Jeffrd10, Diannaa, Minimac, DARTH SIDIOUS 2, Scottcere2k8, Andrea105, Mean as custard, Monkeyboy2601, RjwilmsiBot, Savannah Briggs, Slon02, Zujine, EmausBot, WTM, John of Reading, Orphan Wiki, Sunuraju, Immunize, Super48paul, RA0808, Smoore3, Solarra, Tommy2010, Wikipelli, AsceticRose, Gitghetto, Savh, Ao333, Ularsjya, BurtAlert, Fæ, Shuipzv3, Rppeabody, Iñaki Salazar, Gashrash, Ganesh Paudel, AOC25, MoireL5522, Kiwi128, Alpha Quadrant (alt), Medeis, YaredBTesfaye, 4l31st3r, QEDK, Tolly4bolly, Cmathio, Someone65, Rcsprinter123, Glennconti, Tercerista, Pengkeu, Mohamad yaish, Jacobhankat, Zhanglong~enwiki, L Kensington, Shrigley, Kate0504, Donner60, SBaker43, Xiaoyu of Yuxi, Puffin, Wipsenade, ElockidAlternate, Orange Suede Sofa, Bill william compton, Lizaweb, NTox, Uziel302, DASHBotAV, Zchilds53, Cwek, Marlz666, Pencilcasesrock42, 28bot, Estheroliver, Petrb, Charlie7534, ClueBot NG, Onanoff, Jack Greenmaven, Roaringjohn, Work2win, This lousy T-shirt, Kejia, Satellizer, Joefromrandb, HectorAE, Millermk, SunCountryGuy01, ElenaP1988, Neljack, BlowingTopHat, Widr, Karl 334, Antiqueight, Gao le, Ecade7, Shimmeringstardust, Helpful Pixie Bot, In actu, AzureAnt, Eok2110, Calabe1992, Zybtl, Callumbobbyea, BG19bot, Joshua2278, Engirst, Leedsrule66, Wiki13, MusikAnimal, Darkness Shines, Cold Season, Dan653, Mark Arsten, Silvrous, FutureTrillionaire, John fisher dumass, 4Camelot4, JamesBrown5292, AndreiR16, Tylse123, Klilidiplomus, Dhindley10, Fylbecatulous, Shredder2012, Muffin Wizard, Dtlocm, BattyBot, Biosthmors, Millennium bug, Prof. Squirrel, David.moreno72, ISimonHD, Littlemanks, Danielashdownhighbridge, Chiken'licken, Pratyya Ghosh, Edit1000, Cyberbot II, Ezios Prodigy, HappyJelly, Jionpedia, Halewood37, YFdyh-bot, TheJJJunk, EuroCarGT, Ducknish, 2Flows, Rinkle gorge, Kyleiscool24, SmileyLlama, Hellodarlingbear, Jassy2602, Majiaerhao, Mogism, RazrRekr201, Telstarpk, Fifastar01, Spelling Style, Lugia2453, 54Abby, Reatlas, Killuminator, Vs trush, Epicgenius, Dimitreelz, Marxistfounder, ATListhebest, Hassanmalybob, PrecisionEditorial, Acetotyce, Onechildpolicy, BreakfastJr, Melonkelon, Page00, Surfer43, Sattar91, Everymorning, SPA3142, Xie Luo Fen, Boab McGinlay, AndyDrewYou, TheForumTroller, DavidLeighEllis, CensoredScribe, Babitaarora, Thevideodrome, Cogsdragonfire, Da morrow, NottNott, Finnusertop, Ginsuloft, RainCity471, Zwerg Nase, Param Mudgal, Shavaan, 1jkziminski, Skr15081997, Navar.italia95, Nsmith3190, Joppa Chong, Killer360night, Monkbot, Zumoarirodoka, Chesnaught555, Coolman9999, SantiLak, LOLCOMEATME, GinAndChronically, Reyesvictor, Brittany B., Macofe, Krishna Pagadala, Gcolbert10, Malerisch, Davisonio, InternationalistChap, ABasinger001, Appleangel11, CheeseyFlim, Ur bestie123, Bengoodrick2000, Channelguess, Danielfuinogl, Wikbot15, Waters.Justin, Eteethan, OberyntheViper, Kondom2, Nirmay001, Quivico, Cehn, Karimtaha1997, 东边雨下, Empour, GeneralizationsAreBad, YOWASSUPSOUP, Articlegirl123, JenniferTheEmpress0, Maxman878, Hey Its BOSS, Sregor4, Dutral, 75quidnunc, MarkYabloko, Lachthedrummer19, CLCStudent, Tabiibnafsanii, Winterysteppe, WayfaringWanderer, In veritas, GSS-1987, Slim Slimy, Cabgsddskfbndgfskdfvsnfdgksgfdjsfdgkdvfn, Roomsthinker, Doulph88, H.dryad, Vortexwarrior10, Jzhan119, Pao bhaji, The subliminal Prefect and Anonymous: 1890

- **Tan Kai** *Source:* https://en.wikipedia.org/wiki/Tan_Kai?oldid=707805642 *Contributors:* Edcolins, Gene Nygaard, Sdornan, Badagnani, DarthVader, Hmains, Cydebot, Holderlin, OhanaUnited, SiobhanHansa, Magioladitis, Waacstats, Xue hanyu, Arizonasqueeze, TheSlowLife, DumZiBoT, Yobot, Xiaoyu of Yuxi, KasparBot and Anonymous: 3

- **Wu Lihong** *Source:* https://en.wikipedia.org/wiki/Wu_Lihong?oldid=712784477 *Contributors:* Davidcannon, Alan Liefting, Edcolins, Rich Farmbrough, Vsmith, Stemonitis, Woohookitty, Wavelength, Badagnani, Waacstats, VolkovBot, VonHamburger, Hadoooookin, Addbot, Download, Everweb, Xqbot, RjwilmsiBot, Bagbone, ChiuTaitai, Xiaoyu of Yuxi, KasparBot and Anonymous: 2

7.2 Images

- **File:1_panda_trio_sichuan_china_2011.jpg** *Source:* https://upload.wikimedia.org/wikipedia/commons/1/1a/1_panda_trio_sichuan_china_2011.jpg *License:* GFDL *Contributors:* chensiyuan *Original artist:* chensiyuan

- **File:2007_Coal_Reserves_in_BTUs.png** *Source:* https://upload.wikimedia.org/wikipedia/commons/2/2f/2007_Coal_Reserves_in_BTUs.png *License:* CC BY-SA 3.0 *Contributors:* Own work *Original artist:* Peace01234

- **File:2015_Air_pollution_in_Beijing.svg** *Source:* https://upload.wikimedia.org/wikipedia/commons/7/7a/2015_Air_pollution_in_Beijing.svg *License:* CC BY-SA 4.0 *Contributors:* Own work

Data source: Mission China, Beijing - Historical Data. US Department of States.

- **File:Chinese_animal_002.jpg** *Source:* https://upload.wikimedia.org/wikipedia/commons/3/39/Chinese_animal_002.jpg *License:* CC0 *Contributors:* Own work *Original artist:* Anna Frodesiak
- **File:Chronic_obstructive_pulmonary_disease_world_map_-_DALY_-_WHO2004.svg** *Source:* https://upload.wikimedia.org/wikipedia/commons/b/b8/Chronic_obstructive_pulmonary_disease_world_map_-_DALY_-_WHO2004.svg *License:* CC BY-SA 2.5 *Contributors:*
- Vector map from BlankMap-World6, compact.svg by Canuckguy et al. *Original artist:* Lokal_Profil
- **File:Coal_Bike,_China_1997.jpg** *Source:* https://upload.wikimedia.org/wikipedia/commons/1/1e/Coal_Bike%2C_China_1997.jpg *License:* CC BY-SA 2.0 *Contributors:* 1997 China - Coal Bike *Original artist:* Brian Kelley from Auggen, Germany
- **File:Coal_hopper_with_barge_Rob_Loftis.jpeg** *Source:* https://upload.wikimedia.org/wikipedia/commons/e/eb/Coal_hopper_with_barge_Rob_Loftis.jpeg *License:* CC BY 3.0 *Contributors:*
  Coal hopper with barge.jpg

  Original artist: Rob Loftis
- **File:Coal_mine_in_Inner_Mongolia_002.jpg** *Source:* https://upload.wikimedia.org/wikipedia/commons/0/00/Coal_mine_in_Inner_Mongolia_002.jpg *License:* CC BY 2.0 *Contributors:* Mongolia *Original artist:* Herry Lawford from London, UK
- **File:Commons-logo.svg** *Source:* https://upload.wikimedia.org/wikipedia/en/4/4a/Commons-logo.svg *License:* CC-BY-SA-3.0 *Contributors:* ? *Original artist:* ?
- **File:Crystal_Clear_app_kedit.svg** *Source:* https://upload.wikimedia.org/wikipedia/commons/e/e8/Crystal_Clear_app_kedit.svg *License:* LGPL *Contributors:* Sabine MINICONI *Original artist:* Sabine MINICONI
- **File:Crystal_energy.svg** *Source:* https://upload.wikimedia.org/wikipedia/commons/1/14/Crystal_energy.svg *License:* LGPL *Contributors:* Own work conversion of Image:Crystal_128_energy.png *Original artist:* Dhatfield
- **File:Danshan_Nongguang_Village_Bulletin_board.jpg** *Source:* https://upload.wikimedia.org/wikipedia/commons/d/db/Danshan_Nongguang_Village_Bulletin_board.jpg *License:* Public domain *Contributors:* Own work *Original artist:* David Cowhig
- **File:Dongtan.jpg** *Source:* https://upload.wikimedia.org/wikipedia/commons/4/48/Dongtan.jpg *License:* CC BY 3.0 *Contributors:* Self-photographed *Original artist:* Sh1019
- **File:Dryrivernearbeijing.jpg** *Source:* https://upload.wikimedia.org/wikipedia/commons/e/e2/Dryrivernearbeijing.jpg *License:* CC BY 3.0 *Contributors:* Own work *Original artist:* F3rn4nd0
- **File:Earth_Day_Flag.png** *Source:* https://upload.wikimedia.org/wikipedia/commons/6/6a/Earth_Day_Flag.png *License:* Public domain *Contributors:* File:Earth flag PD.jpg, File:The Earth seen from Apollo 17 with transparent background.png *Original artist:* NASA (Earth photograph) SiBr4 (flag image)
- **File:Edit-clear.svg** *Source:* https://upload.wikimedia.org/wikipedia/en/f/f2/Edit-clear.svg *License:* Public domain *Contributors:* The *Tango!* Desktop Project. *Original artist:*
  The people from the Tango! project. And according to the meta-data in the file, specifically: "Andreas Nilsson, and Jakub Steiner (although minimally)."
- **File:Emblem_of_Regional_Council,_Hong_Kong.svg** *Source:* https://upload.wikimedia.org/wikipedia/commons/3/34/Emblem_of_Regional_Council%2C_Hong_Kong.svg *License:* Public domain *Contributors:* Remade according to official specification *Original artist:* User:Wrightbus (SVG code)
- **File:Factory_in_China.jpg** *Source:* https://upload.wikimedia.org/wikipedia/commons/4/47/Factory_in_China.jpg *License:* CC BY 2.0 de *Contributors:* Own work *Original artist:* High Contrast
- **File:Flag_of_Hong_Kong.svg** *Source:* https://upload.wikimedia.org/wikipedia/commons/5/5b/Flag_of_Hong_Kong.svg *License:* Public domain *Contributors:* http://www.protocol.gov.hk/flags/chi/r_flag/index.html *Original artist:* Tao Ho
- **File:Flag_of_the_People'{}s_Republic_of_China.svg** *Source:* https://upload.wikimedia.org/wikipedia/commons/f/fa/Flag_of_the_People%27s_Republic_of_China.svg *License:* Public domain *Contributors:* Own work, http://www.protocol.gov.hk/flags/eng/n_flag/design.html *Original artist:* Drawn by User:SKopp, redrawn by User:Denelson83 and User:Zscout370
- **File:Folder_Hexagonal_Icon.svg** *Source:* https://upload.wikimedia.org/wikipedia/en/4/48/Folder_Hexagonal_Icon.svg *License:* Cc-by-sa-3.0 *Contributors:* ? *Original artist:* ?
- **File:Food_and_Environmental_Hygiene_Department_Logo.svg** *Source:* https://upload.wikimedia.org/wikipedia/en/5/5c/Food_and_Environmental_Hygiene_Department_Logo.svg *License:* Fair use *Contributors:*
  www.fehd.gov.hk/english/cc/die_todo_e.pdf
  Original artist: ?
- **File:Global_Warming_Map.jpg** *Source:* https://upload.wikimedia.org/wikipedia/commons/8/8c/Global_Warming_Map.jpg *License:* CC-BY-SA-3.0 *Contributors:* ? *Original artist:* ?
- **File:Gobi_desert_en.jpg** *Source:* https://upload.wikimedia.org/wikipedia/commons/5/5b/Gobi_desert_en.jpg *License:* CC BY-SA 3.0 *Contributors:* Own work *Original artist:* Christophe cagé
- **File:Green_pog.svg** *Source:* https://upload.wikimedia.org/wikipedia/commons/a/ab/Green_pog.svg *License:* Public domain *Contributors:* ? *Original artist:* ?
- **File:Guiyu-ewaste.jpg** *Source:* https://upload.wikimedia.org/wikipedia/commons/0/07/Guiyu-ewaste.jpg *License:* CC BY-SA 3.0 *Contributors:* Own work *Original artist:* bleahbleahbleah

- **File:Question_book-new.svg** *Source:* https://upload.wikimedia.org/wikipedia/en/9/99/Question_book-new.svg *License:* Cc-by-sa-3.0 *Contributors:*
Created from scratch in Adobe Illustrator. Based on Image:Question book.png created by User:Equazcion *Original artist:*
Tkgd2007

- **File:Rambler_Channel_1.jpg** *Source:* https://upload.wikimedia.org/wikipedia/commons/5/5b/Rambler_Channel_1.jpg *License:* GFDL *Contributors:* Own work *Original artist:* Minghong

- **File:Red_pog.svg** *Source:* https://upload.wikimedia.org/wikipedia/en/0/0c/Red_pog.svg *License:* Public domain *Contributors:* ? *Original artist:* ?

- **File:SaiTsoWan_Decoration_Toilet.jpg** *Source:* https://upload.wikimedia.org/wikipedia/commons/f/f5/SaiTsoWan_Decoration_Toilet.jpg *License:* CC BY-SA 3.0 *Contributors:* Own work *Original artist:* Ngchikit

- **File:Sai_Tso_Wan.jpg** *Source:* https://upload.wikimedia.org/wikipedia/commons/2/2e/Sai_Tso_Wan.jpg *License:* CC BY-SA 2.5 *Contributors:* Own work *Original artist:* me

- **File:Sex_ratio_at_birth_in_mainland_China.png** *Source:* https://upload.wikimedia.org/wikipedia/commons/a/ae/Sex_ratio_at_birth_in_mainland_China.png *License:* Public domain *Contributors:* "China's unbalanced sex ratio at birth, millions of excess bachelors and societal implications" *Vulnerable Children and Youth Studies* **6**(4):314-20 doi:10.1080/17450128.2011.630428 *Original artist:* Poston, D.L., Jr., *et al*

- **File:ShaTinSewageTreatmentWorks_BirdEyeView_2.jpg** *Source:* https://upload.wikimedia.org/wikipedia/commons/c/cb/ShaTinSewageTreatmentWorks_BirdEyeView_2.jpg *License:* Public domain *Contributors:* Own work (Self took photo) *Original artist:* Chong Fat

- **File:Sha_Tin_Sewage_Treatment_Works.jpg** *Source:* https://upload.wikimedia.org/wikipedia/commons/0/06/Sha_Tin_Sewage_Treatment_Works.jpg *License:* CC BY-SA 3.0 *Contributors:* Own work *Original artist:* WiNG

- **File:Sidings_and_shaft_entry.jpg** *Source:* https://upload.wikimedia.org/wikipedia/commons/2/29/Sidings_and_shaft_entry.jpg *License:* CC BY-SA 2.0 *Contributors:* Sidings and shaft entry *Original artist:* Peter Van den Bossche from Mechelen, Belgium

- **File:Sin_Fat_Road,_Sai_Tso_Wan.jpg** *Source:* https://upload.wikimedia.org/wikipedia/commons/5/55/Sin_Fat_Road%2C_Sai_Tso_Wan.jpg *License:* CC BY-SA 2.5 *Contributors:* Own work *Original artist:* Deryck Chan

- **File:Tieshan-solar-water-heaters-0101.jpg** *Source:* https://upload.wikimedia.org/wikipedia/commons/9/92/Tieshan-solar-water-heaters-0101.jpg *License:* CC-BY-SA-3.0 *Contributors:* Own work (own photo) *Original artist:* User:Vmenkov

- **File:Unbalanced_scales.svg** *Source:* https://upload.wikimedia.org/wikipedia/commons/f/fe/Unbalanced_scales.svg *License:* Public domain *Contributors:* ? *Original artist:* ?

- **File:Urban_Council_(Coat_of_Arms).jpg** *Source:* https://upload.wikimedia.org/wikipedia/commons/f/f5/Urban_Council_%28Coat_of_Arms%29.jpg *License:* CC BY-SA 3.0 *Contributors:* Own work *Original artist:* Clithering

- **File:Urban_Council_Public_Libraries_Reading_Programme.jpg** *Source:* https://upload.wikimedia.org/wikipedia/en/f/f3/Urban_Council_Public_Libraries_Reading_Programme.jpg *License:* Cc-by-sa-3.0 *Contributors:* ? *Original artist:* ?

- **File:Wiki_letter_w.svg** *Source:* https://upload.wikimedia.org/wikipedia/en/6/6c/Wiki_letter_w.svg *License:* Cc-by-sa-3.0 *Contributors:* ? *Original artist:* ?

- **File:Wiki_letter_w_cropped.svg** *Source:* https://upload.wikimedia.org/wikipedia/commons/1/1c/Wiki_letter_w_cropped.svg *License:* CC-BY-SA-3.0 *Contributors:* This file was derived from Wiki letter w.svg: <a href='//commons.wikimedia.org/wiki/File:Wiki_letter_w.svg' class='image'><img alt='Wiki letter w.svg' src='https://upload.wikimedia.org/wikipedia/commons/thumb/6/6c/Wiki_letter_w.svg/50px-Wiki_letter_w.svg.png' width='50' height='50' srcset='https://upload.wikimedia.org/wikipedia/commons/thumb/6/6c/Wiki_letter_w.svg/75px-Wiki_letter_w.svg.png 1.5x, https://upload.wikimedia.org/wikipedia/commons/thumb/6/6c/Wiki_letter_w.svg/100px-Wiki_letter_w.svg.png 2x' data-file-width='44' data-file-height='44' /></a>
Original artist: Derivative work by Thumperward

- **File:Wikinews-logo.svg** *Source:* https://upload.wikimedia.org/wikipedia/commons/2/24/Wikinews-logo.svg *License:* CC BY-SA 3.0 *Contributors:* This is a cropped version of Image:Wikinews-logo-en.png. *Original artist:* Vectorized by Simon 01:05, 2 August 2006 (UTC) Updated by Time3000 17 April 2007 to use official Wikinews colours and appear correctly on dark backgrounds. Originally uploaded by Simon.

- **File:Wind-turbine-icon.svg** *Source:* https://upload.wikimedia.org/wikipedia/commons/a/ad/Wind-turbine-icon.svg *License:* CC BY-SA 3.0 *Contributors:* Own work *Original artist:* Lukipuk

- **File:Yangzhou-WenchangGe-traffic-3417.jpg** *Source:* https://upload.wikimedia.org/wikipedia/commons/b/bd/Yangzhou-WenchangGe-traffic-3417.jpg *License:* CC BY-SA 3.0 *Contributors:* Own work *Original artist:* Vmenkov

- **File:Yangzhou-WenchangLu-electric-bicycles-3278.jpg** *Source:* https://upload.wikimedia.org/wikipedia/commons/c/c0/Yangzhou-WenchangLu-electric-bicycles-3278.jpg *License:* CC BY-SA 3.0 *Contributors:* Own work *Original artist:* Vmenkov

- **File:Zhongwen.svg** *Source:* https://upload.wikimedia.org/wikipedia/commons/9/9e/Zhongwen.svg *License:* Public domain *Contributors:* ? *Original artist:* ?

7.3 Content license